AF321588

THE LAST
TEN YEARS

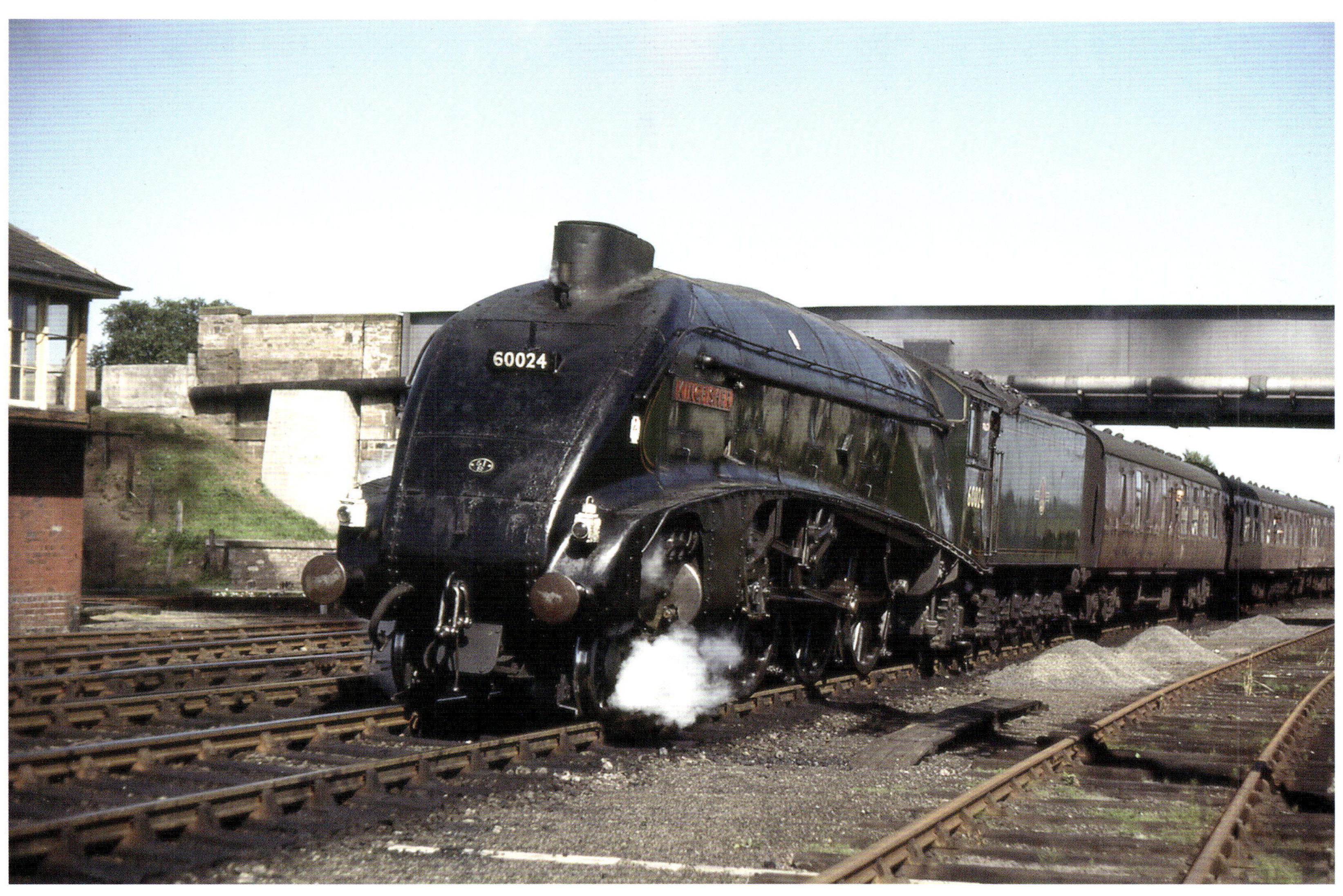

The History Press books by the same author:

Steam in Scotland: The Railway Photographs of R.J. (Ron) Buckley

Southern Steam: The Railway Photographs of R.J. (Ron) Buckley

Steam in the North East: The Railway Photographs of R.J. (Ron) Buckley

Great Western Steam: The Railway Photographs of R.J. (Ron) Buckley

Steam in the East Midlands and East Anglia: The Railway Photographs of R.J. (Ron) Buckley

London Midland Steam: The Railway Photographs of R.J. (Ron) Buckley

Early and First Generation Green Diesels in Photographs

Steam in the 1950s: The Railway Photographs of Robert Butterfield

Scottish Steam 1948–1966: The Railway Photographs of Andrew Grant Forsyth

East and North Eastern Steam 1947–1958: The Railway Photographs of Andrew Grant Forsyth

Diesel Power in the North Eastern Region
Scottish Steam 1948–1967

THE LAST TEN YEARS

Colour Photographs of the End of Steam, 1959–68

Brian J. Dickson

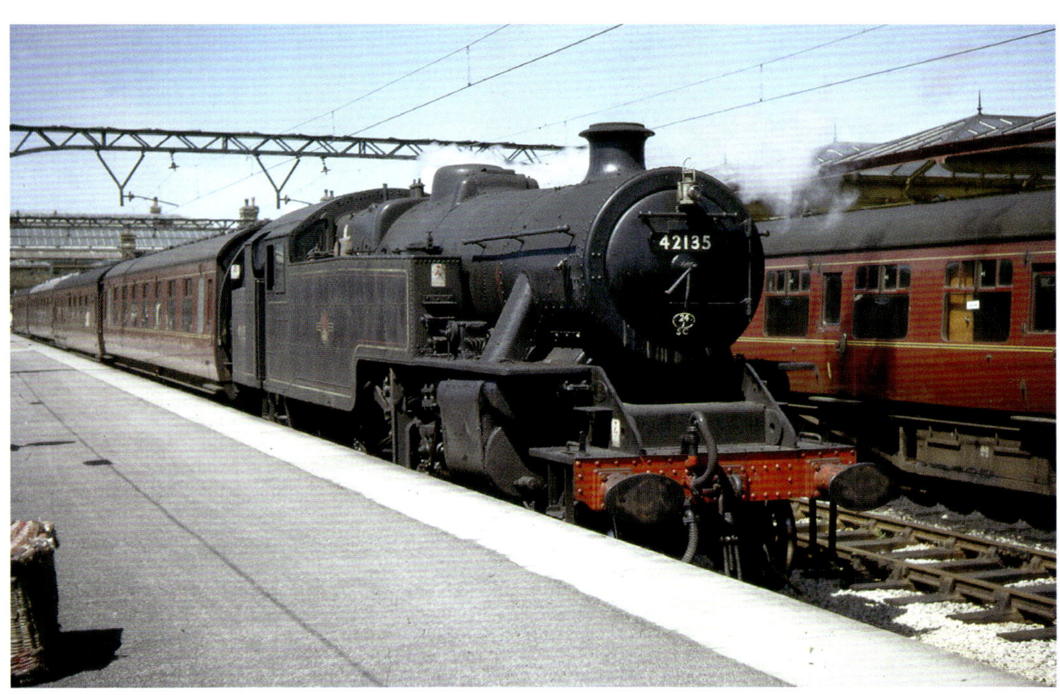

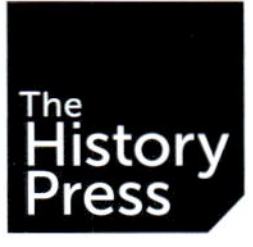

Front cover: Sunday 11 February 1962. Ex-LNER Class A3 4-6-2 No. 60061 *Pretty Polly* is seen at Kings Cross shed in London. (Hugh D. Ramsay)

Back cover: Thursday 28 March 1968. Seen lifting a train of loaded coal wagons away from a stand near Rose Grove is ex-LMS Class 8F 2-8-0 No. 48247. Originally constructed by the NBL as part of an order from the Ministry of Supply during 1940, she would spend time serving in Persia before returning to the UK and being purchased by British Railways in 1949. She would be withdrawn from service later in 1968. (Andrew G. Forsyth)

Half-title page: Thursday 25 August 1966. Ex-LNER Class A4 4-6-2 No. 60024 *Kingfisher* is seen here departing from Bridge of Dun at the head of the 'up' 'Grampian' express from Aberdeen to Glasgow. Even at this late date the Ferryhill staff were still turning out the remaining A4s in sparkling condition. Just over one year later the whole line between Kinnaber Junction and Stanley Junction would be closed to passenger traffic. (Andrew G. Forsyth).

Title page: Sunday 20 May 1962. In comparatively clean condition, Class 4P 2-6-4 tank No. 42135 is seen at Morecambe station. A product of Derby Works, in 1950 she was allocated new to Lancaster Green Ayre shed whose 24J shed code she is carrying. She would end her days based at Wigan, being withdrawn during 1964. (Hugh D. Ramsay).

First published 2022

The History Press
97 St George's Place,
Cheltenham GL50 3QB
www.thehistorypress.co.uk

British Library Cataloguing in Publication Data.
A catalogue record for this book is available from the British Library.

ISBN 978 1 80399 100 9

Typesetting and origination by The History Press
Printed in Turkey by Imak

THE PHOTOGRAPHERS

Anthony Scarsbrook was born in London in 1932 and lived in the New Southgate area. With New Southgate being the closest station to his home, he would spend many hours there joining other enthusiasts and photographers witnessing the steam workings on the East Coast Main Line. Friends during those days included Andrew Forsyth and Hugh Ramsay.

Having spent his career in the banking industry, he retired and moved to Milton Keynes in 1986 and, having been a member of the Railway Correspondence and Travel Society (RCTS) since 1947, he volunteered to assist with the organisation of a new local branch in 1999. He was also a member of the Gresley Society. Anthony died in May 2010.

His collection of colour photographs only extends to approximately 250 transparencies from the mid to late 1960s, but includes some excellent shots which are included in this book.

Andrew Grant Forsyth was born in north London in 1923, and lived and worked in the same area all his life. A lifelong railway enthusiast, he was very much a follower of all things associated with the London and North Eastern Railway, and the greater part of his photographic collection contains images relating to that company and its pre-grouping constituents.

In his spare time Andrew was one of the organisers of the Seafield Railway Club, which arranged visits to sheds and railway sites throughout eastern England and Scotland. The club also published a regular magazine titled *The Locomotive Post*, containing observations from contributing enthusiasts and detailing locomotive stock changes and movements that were taking place throughout the industry. Andrew died at the age of 83.

His collection of colour photographs extends to about 800 transparencies taken from 1959 until the end of main-line steam in the UK in 1968.

Hugh Donald Ramsay was born in Essex in 1935 and worked in the print trade in London as a compositor, later moving to Dunstable and working there. He was editor of a series of magazines published between 1980 and 1985 entitled *Railway Reflections* and author of a book titled *Capital Steam*, published during 1989. Hugh died in 2014 in Dunstable.

His collection of colour photographs consists of about 800 transparencies dating from the early to mid 1960s, including numerous high-quality shots of which many are included in this book.

Fully catalogued by Initial Photographics, copies of the listings can be obtained by writing to:
25 The Limes, Stony Stratford, MK11 1ET.

INTRODUCTION

The last ten years of main-line steam traction in the Scottish, North Eastern, Eastern and London Midland regions saw the increasing speed of withdrawal of older pre-grouping classes, together with many of the 'Big Four' locomotives, as more British Railways Standard locomotives became available. From the later 1950s, the management attempt to quickly bring diesel locomotives into traffic faltered, as many classes being delivered proved unreliable, necessitating the continued use of steam traction. This meant that during the early 1960s much passenger traffic on the East Coast Main Line was still in the hands of Nigel Gresley's Pacifics, with William Stanier's Princess Royal and Princess Coronation locomotives hauling the same traffic on the West Coast Main Line. It was not until mid 1963 that Kings Cross shed in London and Haymarket shed in Edinburgh were closed to steam traction, with diesel power taking over much of this work using the more reliable English Electric Type 4s and powerful new Deltics. One fortunate result of these closures was the transfer from 1962 of no less than fourteen Class A4s to Ferryhill shed in Aberdeen, where they were tasked with hauling the three-hour scheduled passenger trains between Aberdeen and Glasgow Buchanan Street, which they continued to do until the last member of the class, No. 60024 *Kingfisher*, was withdrawn in September 1966. A year later the 45 miles of the 'Strathmore' route between Stanley Junction and Kinnaber Junction would be closed to passenger traffic.

The West Coast Main Line also saw much change, with English Electric Type 4 diesels handling much of the passenger traffic, which subsequently led to the withdrawal of all the Princess Royal class by the end of 1962 and the Princess Coronation locomotives by the end of 1964. With the electrification of the route and its opening between London Euston and Crewe in 1965, steam locomotives were forbidden south of Crewe and many subsequently had yellow diagonal stripes painted on their cab sides to indicate as such.

During this period most goods traffic remained under steam power, with many ex-LNER Pacifics, displaced from passenger work, seen hauling fast goods trains; meanwhile, on former LMS routes the Stanier Black 5s and 8Fs still handled much of this traffic. Steam locomotive power was still very much at the fore in the handling of the mineral traffic in the areas covered in this book, with the coal trains in Fife and the Lothians of Scotland being almost exclusively steam hauled by former North British Railway and LNER 0-6-0s, and the same traffic coming from the collieries in the north-east of England being in the hands of former North Eastern Railway 0-6-0 and 0-8-0 locomotives. The Tyne Dock to Consett iron-ore traffic remained steam hauled by a batch of ten Standard Class 9F 2-10-0s until the end of 1966.

The use of steam traction in Scotland would come to an end in June 1967 with the closure of Thornton Junction and Dunfermline sheds and the withdrawal of the last two main-line steam locomotives that remained working in Scotland. Similarly, in the north-east of England the use of steam traction would end with the closure of Sunderland and Tyne Dock sheds in September of the same year.

Meanwhile, several Lancashire sheds were still supplying steam power for goods traffic in that area, with those at Lostock Hall, Rose Grove and Carnforth busy servicing the remaining Black 5, 8F and Standard locomotives still in use. Closure would come to those sheds in early August 1968.

The photographs in this book are laid out in chronological order to assist the reader in following the run-down of steam power in the years covered, and they show such highlights as the sparkling Class A3s, A4s, Jubilees and Black 5s working both scheduled passenger services and some special workings culminating in the last main-line steam-hauled working on Sunday 11 August 1968. During these years I doubt there were many low points in a steam enthusiast's day, even with filthy, work-stained, badly maintained locomotives seen and heard working on both main-line and branch-line duties; the sight and sound of a steam locomotive working would give pleasure to both the spectator and the photographer.

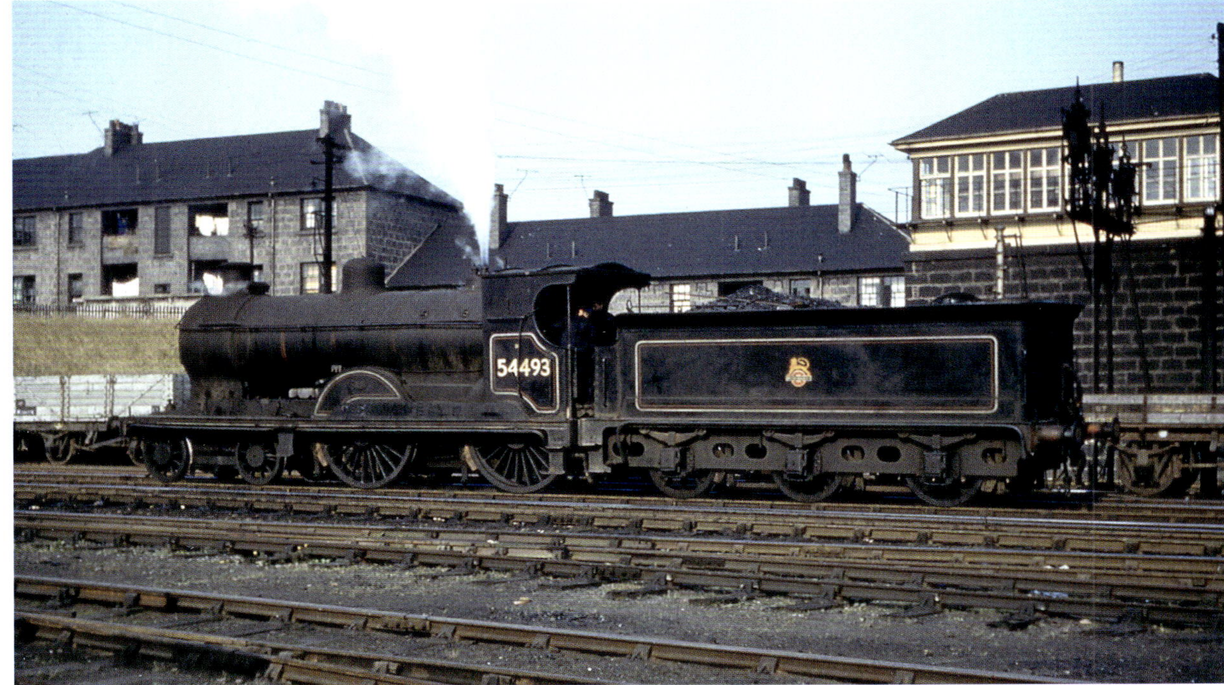

Seen here is a comparison of two of the major Scottish company pre-grouping classes of express passenger 4-4-0s.

Top: Thursday 20 August 1959. Ex-CR Class 113 (LMS Class 3P) No. 54493 standing in the goods yard at Kittybrewster in Aberdeen. Constructed by Armstrong Whitworth & Co. during 1921, she would be withdrawn from service forty years later in 1961. (Andrew G. Forsyth)

Bottom: Friday 21 August 1959. Ex-NBR Class J Superheated Scott (LNER Class D30) No. 62418 *The Pirate* has just been withdrawn from service and is standing in the yard at Thornton Junction shed. Entering service in 1914 from Cowlairs Works, she had been named after the main character in the 1821 published novel of the same name by Sir Walter Scott. (Andrew G. Forsyth)

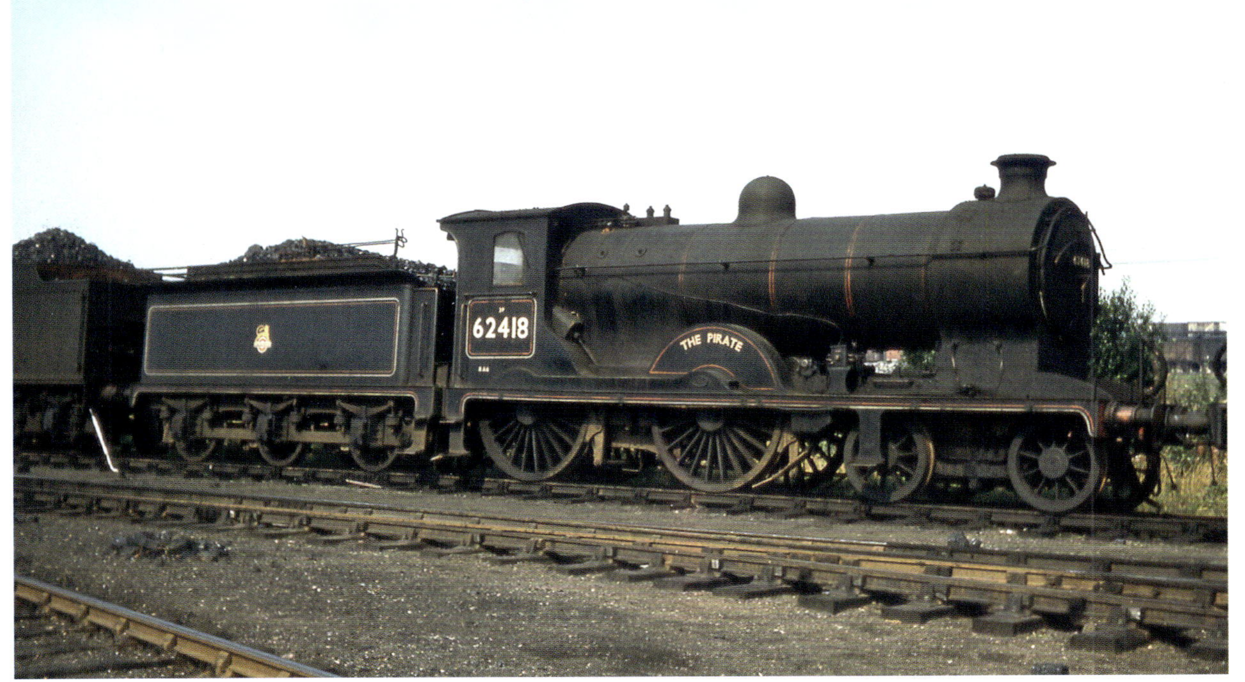

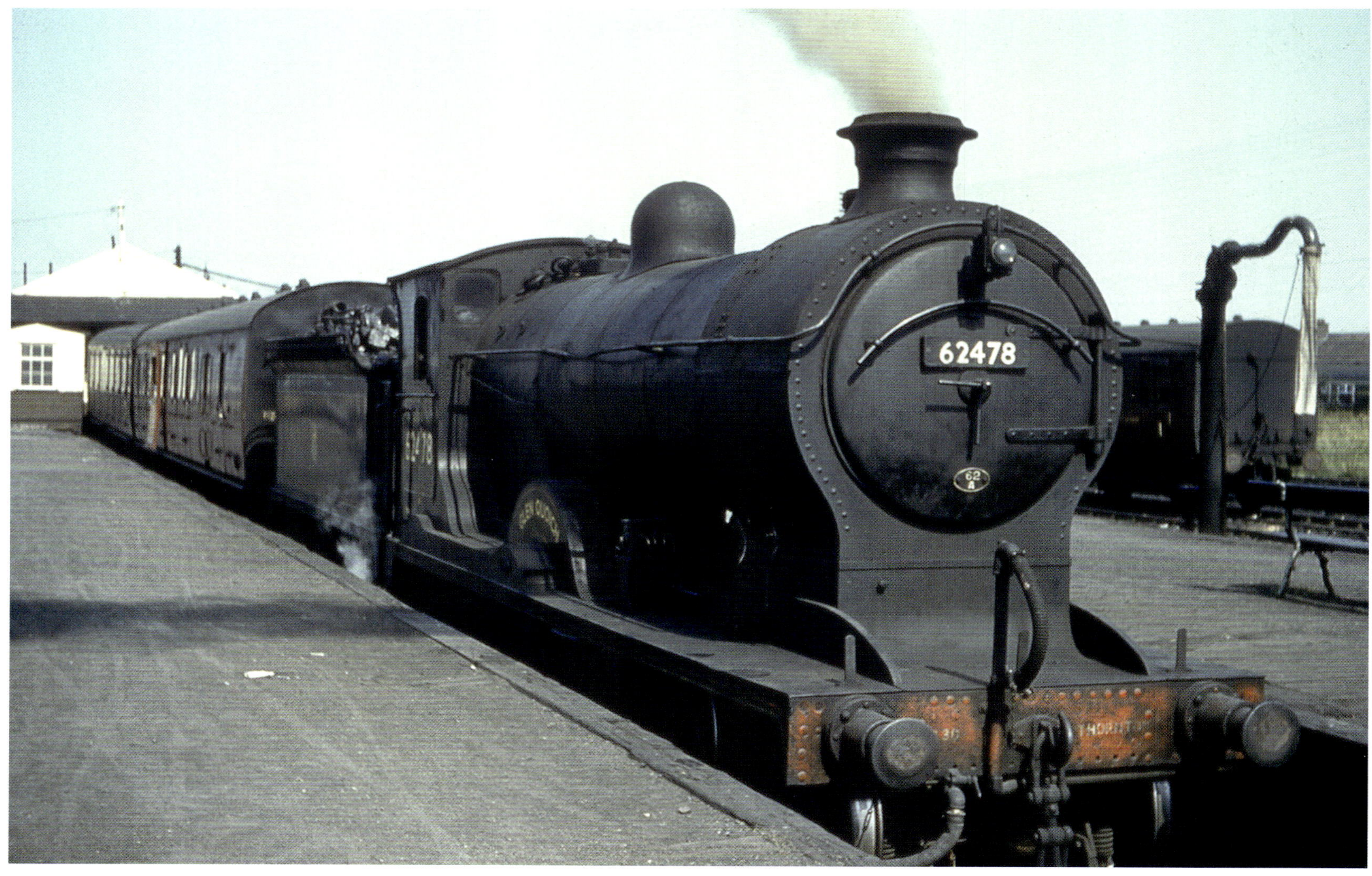

Above: Friday 21 August 1959. Waiting to depart with its two-coach stopper from the 'up' bay platform at Thornton Junction station is ex-NBR Class K (LNER Class D34) Glen 4-4-0 No. 62478 *Glen Quoich*. Entering service from Cowlairs Works during 1917, she would be withdrawn from service four months after this photograph was taken, in December 1959. (Andrew G. Forsyth)

Opposite top: Friday 21 August 1959. Standing outside Dunfermline shed is ex-NBR Class S (LNER Class J37) 0-6-0 No. 64597. Constructed by the NBL in 1919, she would be withdrawn from service during 1961. (Andrew G. Forsyth)

Opposite bottom: Wednesday 26 August 1959. Standing adjacent to the coaling stage at West Hartlepool shed is ex-NER Class P3 (LNER Class J27) 0-6-0 No. 65818. A product of the NBL during 1908, she would give fifty-four years of service before being withdrawn in 1962. (Andrew G. Forsyth)

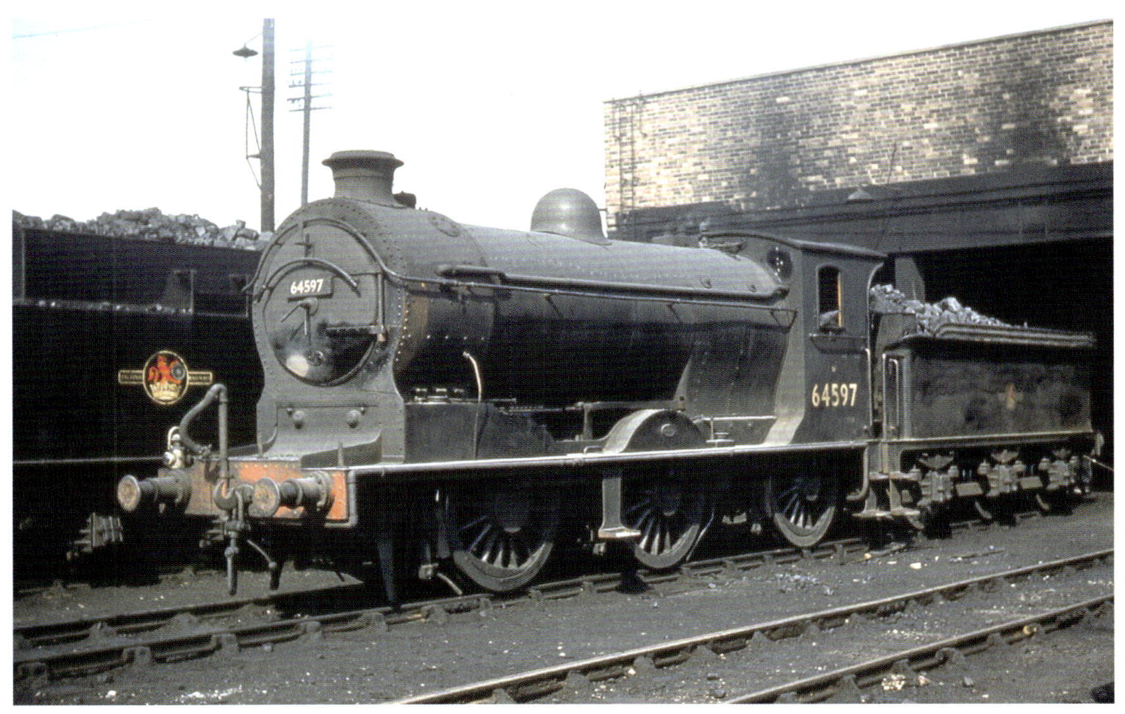

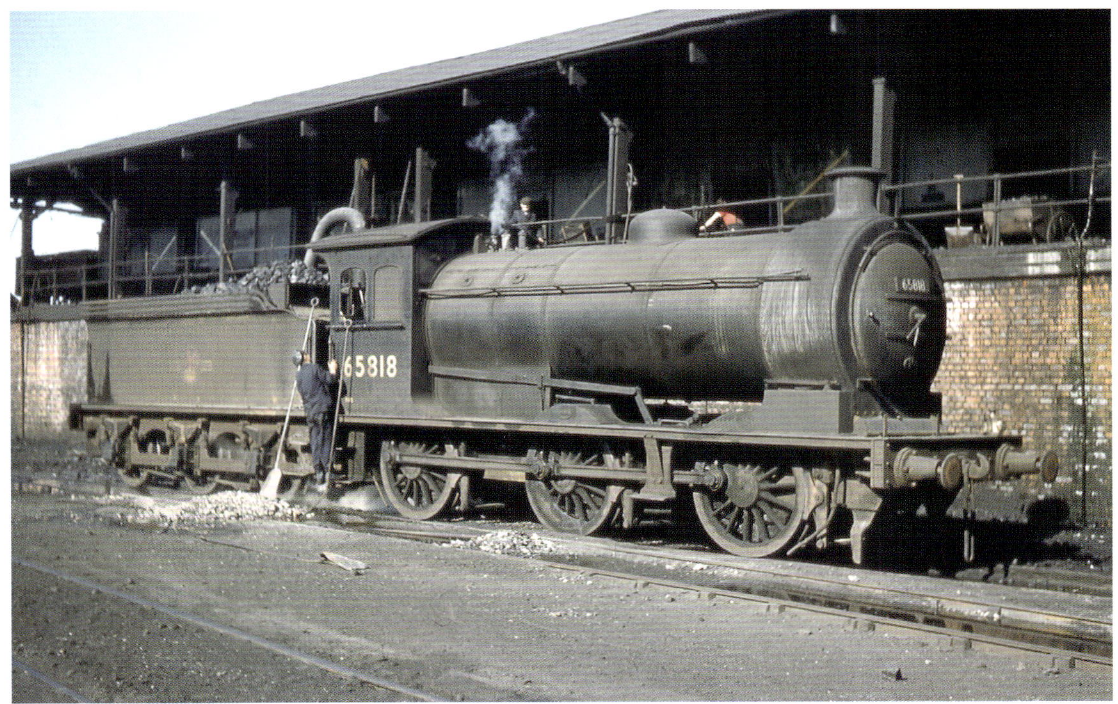

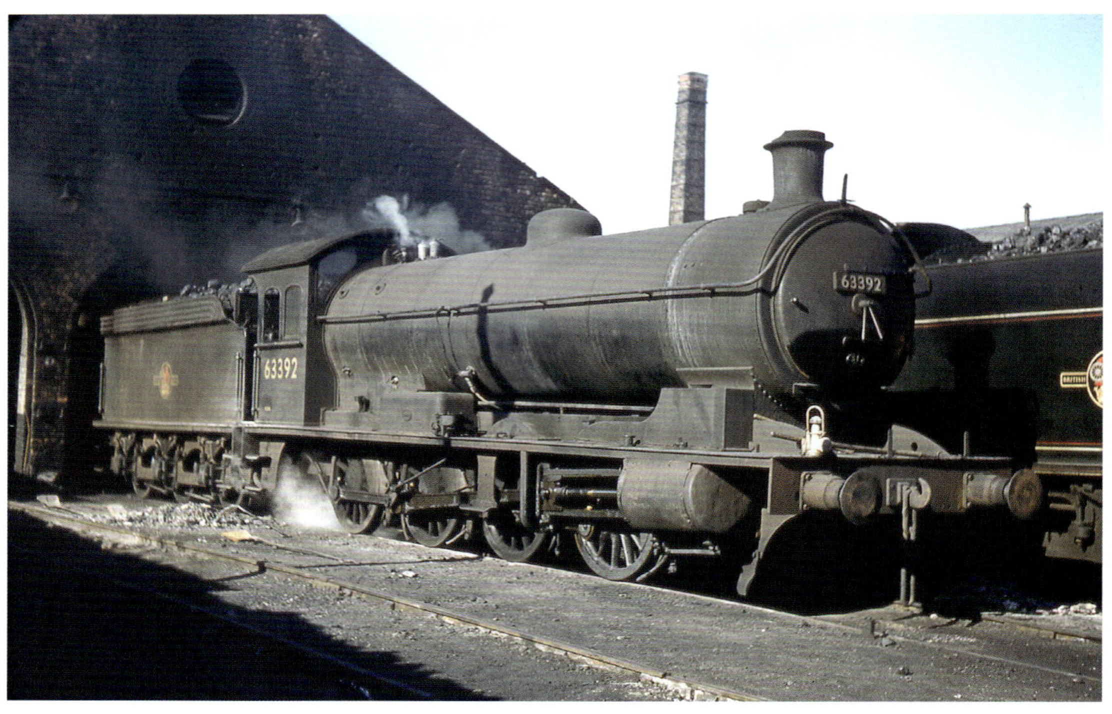

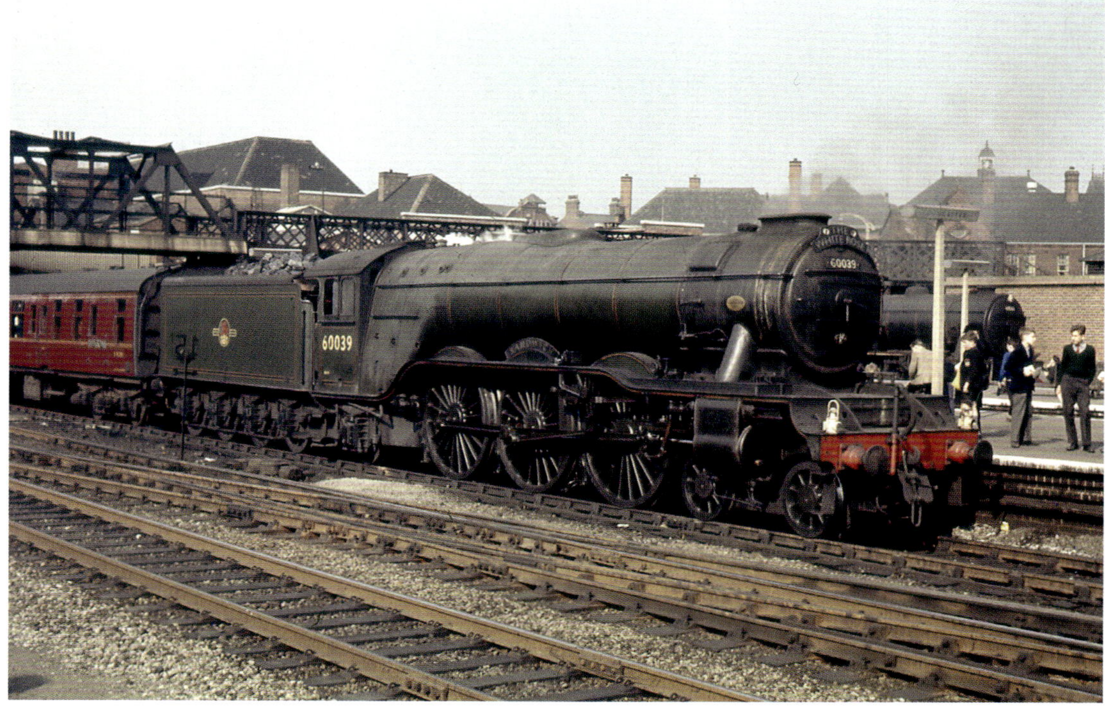

Opposite top: Wednesday 26 August 1959. Also seen at West Hartlepool shed is ex-NER Class T2 (LNER Class Q6) 0-8-0 No. 63392. Entering service from Darlington Works during 1918, she would be withdrawn in 1963. (Andrew G. Forsyth)

Opposite bottom: Wednesday 20 April 1960. Pausing at Doncaster station at the head of an 'up' express is ex-LNER Class A3 4-6-2 No. 60039 *Sandwich*. Constructed at Doncaster Works during 1934 and named after the racehorse that won the 1931 St Leger, she acquired her double chimney in 1959. Fitted with trough-style smoke deflectors during 1961, she would be withdrawn from service in 1963. (Andrew G. Forsyth)

Right: Sunday 14 August 1960. Constructed at Swindon Works during 1957, BR Standard Class 9F 2-10-0 No. 92179 had almost the shortest working life of any of this class, being withdrawn only eight years later in 1965. Seen here at Blaydon shed, west of Newcastle, she is bearing a 34E New England shed code. (Andrew G. Forsyth)

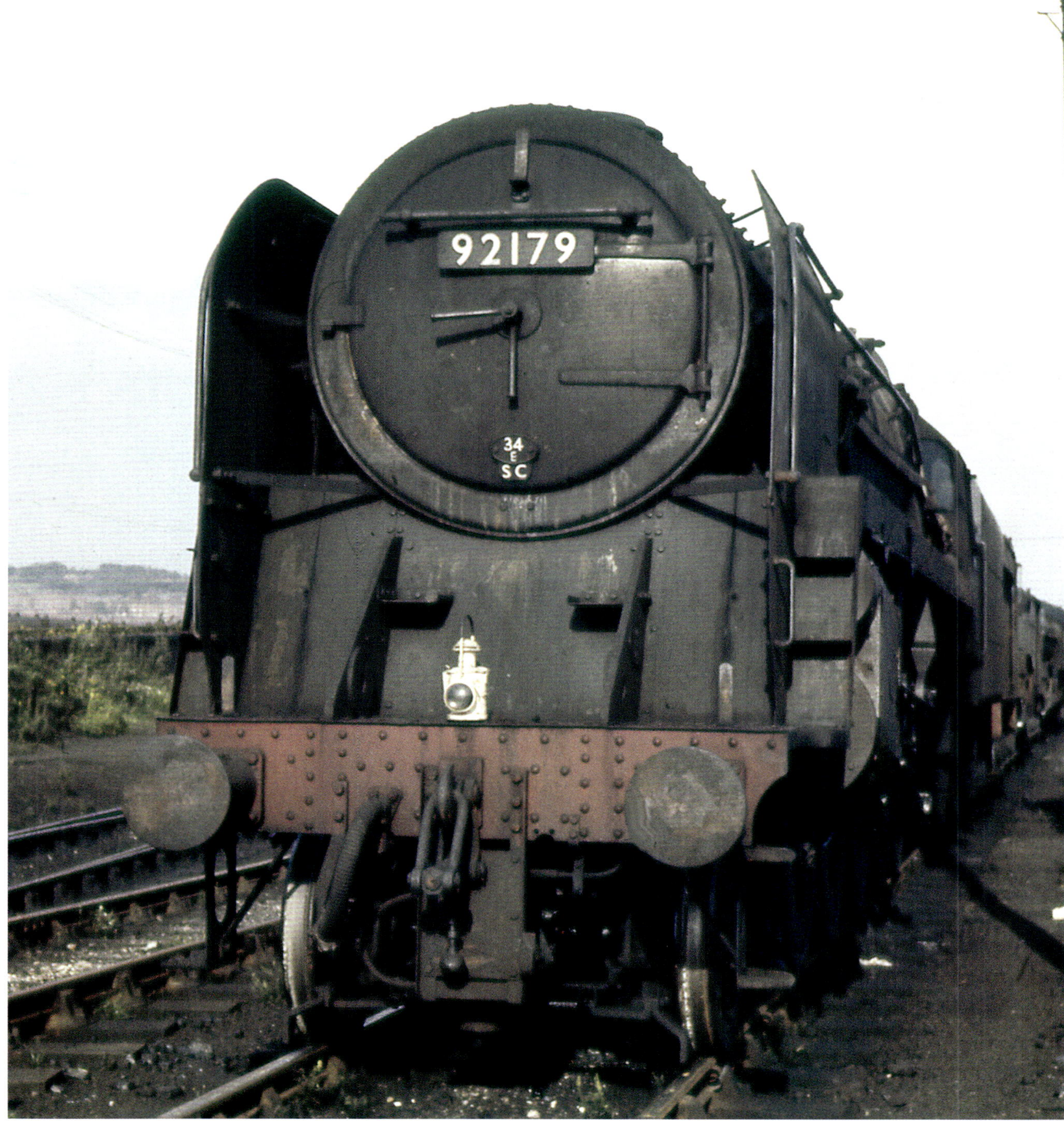

Right: Friday 19 August 1960. Seen departing from platform ten at Edinburgh Waverley station is the summer-only 'The Elizabethan', with ex-LNER Class A4 4-6-2 No. 60032 *Gannet* in charge. Bearing a 34A Kings Cross shed code, the staff there have kept her in first-class condition; note the burnished buffer heads and shackle. Constructed at Doncaster Works in 1938, she would be withdrawn from service in 1963. (Andrew G. Forsyth)

Opposite top: Sunday 26 March 1961. This scene shows Bescot shed 21B, situated south of Walsall on the former L&NWR line. There are at least fourteen locomotives parked, all in steam, awaiting the start of duties on Monday morning. The bulk are ex-LMS Class 4F 0-6-0s with two Class 7F 0-8-0s in the foreground. The shed closed during March 1966 and a new diesel depot was constructed on the site. (Andrew G. Forsyth)

Opposite bottom: Sunday 26 March 1961. Also at Bescot shed is ex-MR Class 1142 (LMS Class 2F) 0-6-0 No. 58122. A real veteran, she was constructed by Kitson & Co. during 1875 and would give eighty-six years of service before being withdrawn later in 1961. (Andrew G. Forsyth)

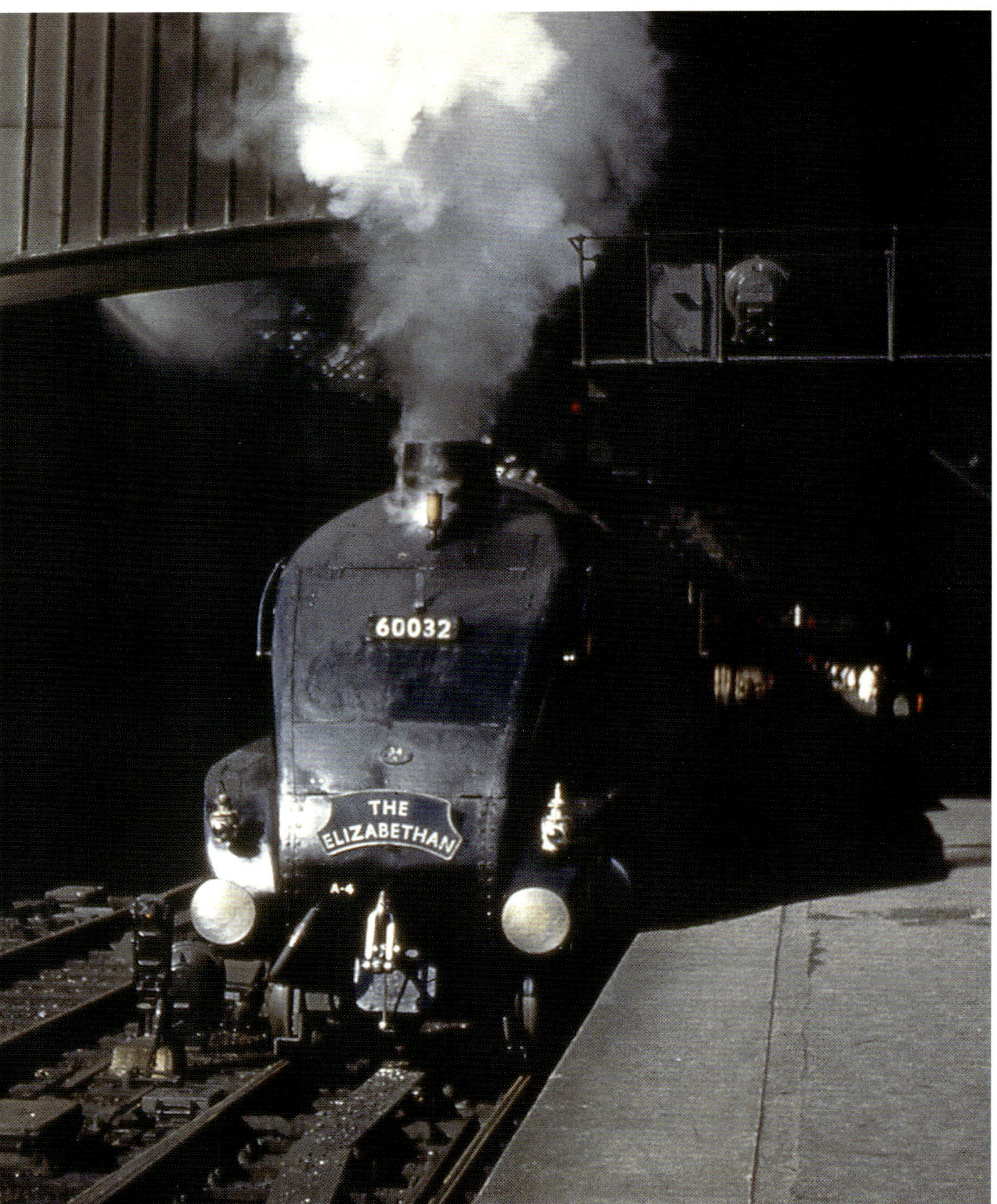

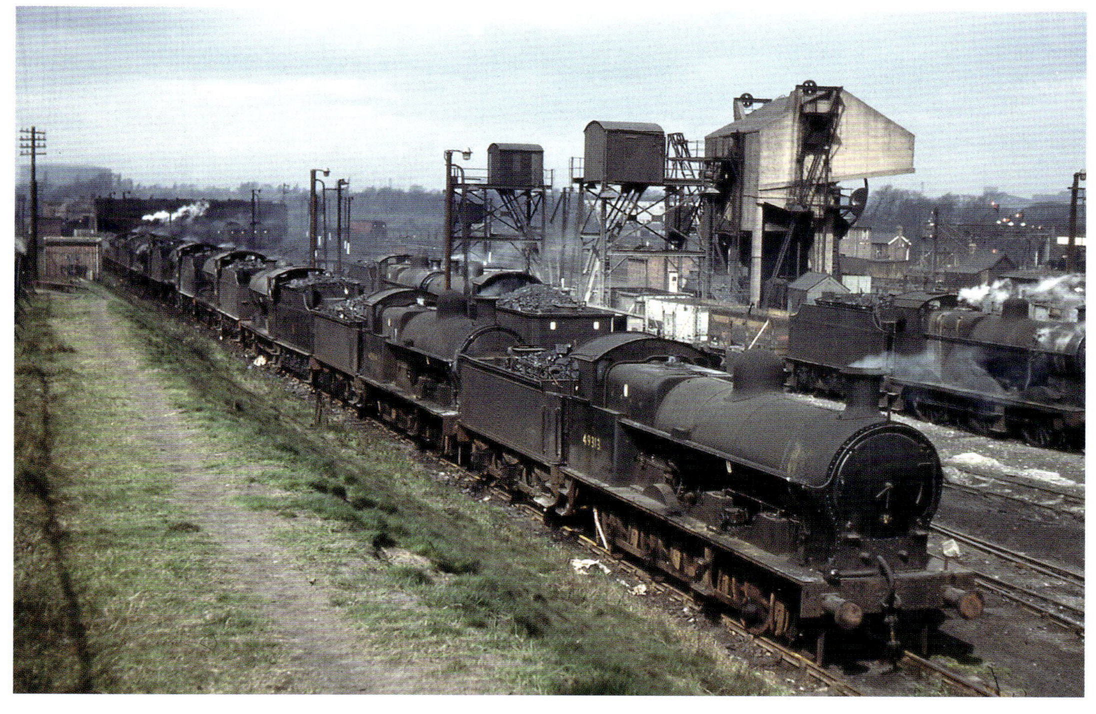

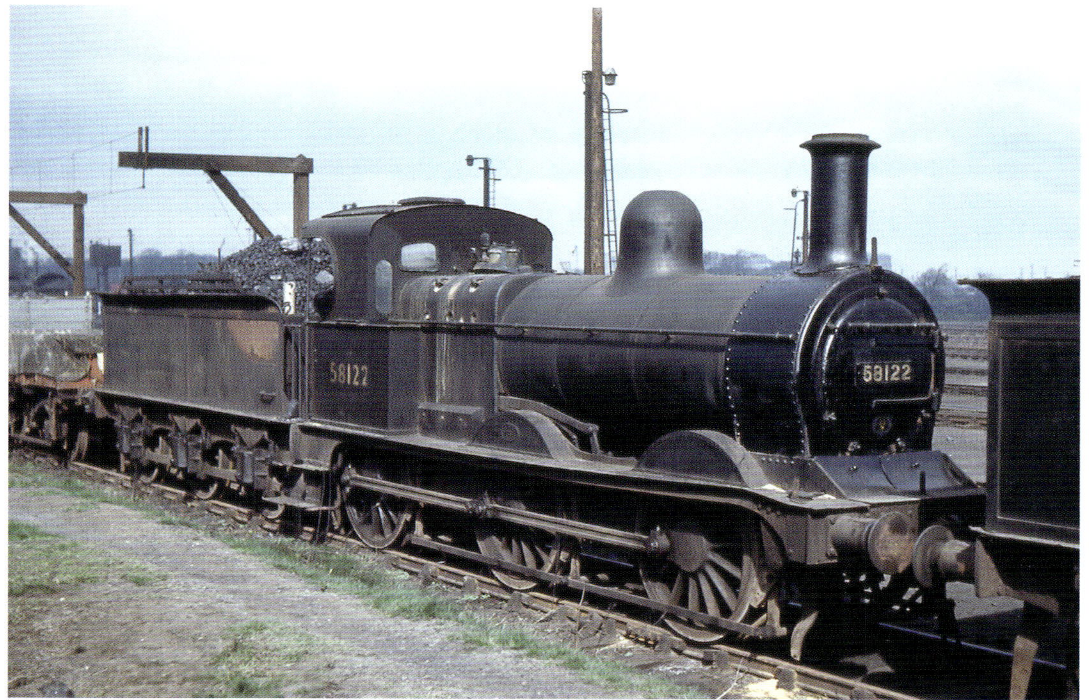

Sunday 9 July 1961. This day saw the running of the RCTS West Riding Branch 'Borders Rail Tour' commencing and returning to Leeds City. Motive power for the day included ex-NBR Class B (LNER Class J37) 0-6-0 No. 64624 and preserved ex-NBR Class K (LNER Class D34) Glen 4-4-0 No. 256 *Glen Douglas*.

Right: Seen in the yard at Hawick shed is No. 64624 waiting to join its train for the journey to Jedburgh and Roxburgh. Constructed by the NBL in 1921, she would be one of the last of the class to be withdrawn in 1966. (Hugh D. Ramsay)

Opposite top: At St Boswells, No. 256, coupled to No. 64624, is seen running round the train. Constructed at Cowlairs Works in 1913, she was withdrawn for preservation during August 1959 and returned for special duties in the NBR livery. She is currently to be seen at the Riverside Museum in Glasgow. (Hugh D. Ramsay)

Opposite bottom: Monday 7 August 1961. At Bridlington shed, Class B1 4-6-0 No. 61303 is bearing a 51L Thornaby shed code. Constructed by the NBL in 1948, she would be withdrawn during 1966. (Hugh D. Ramsay)

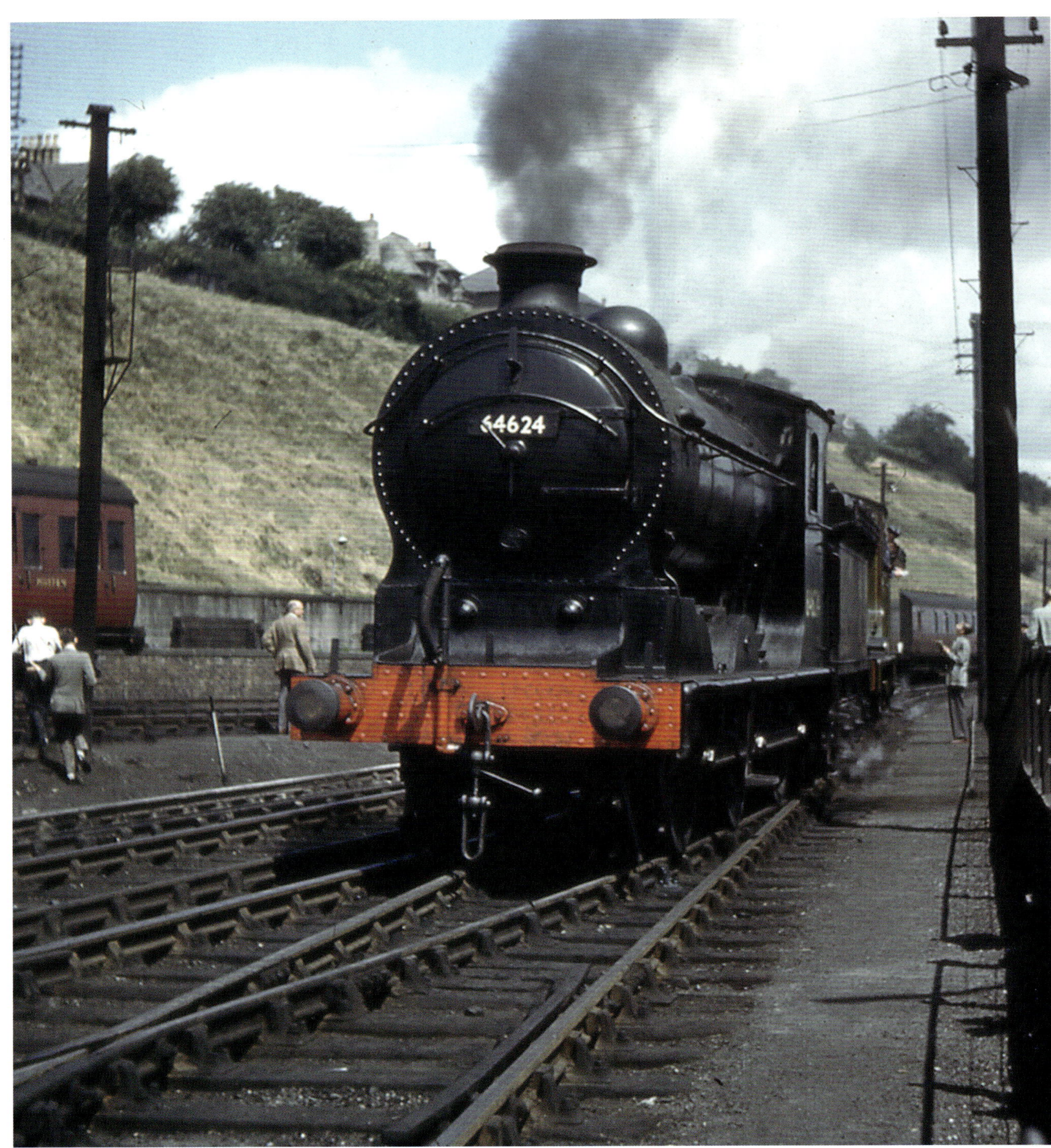

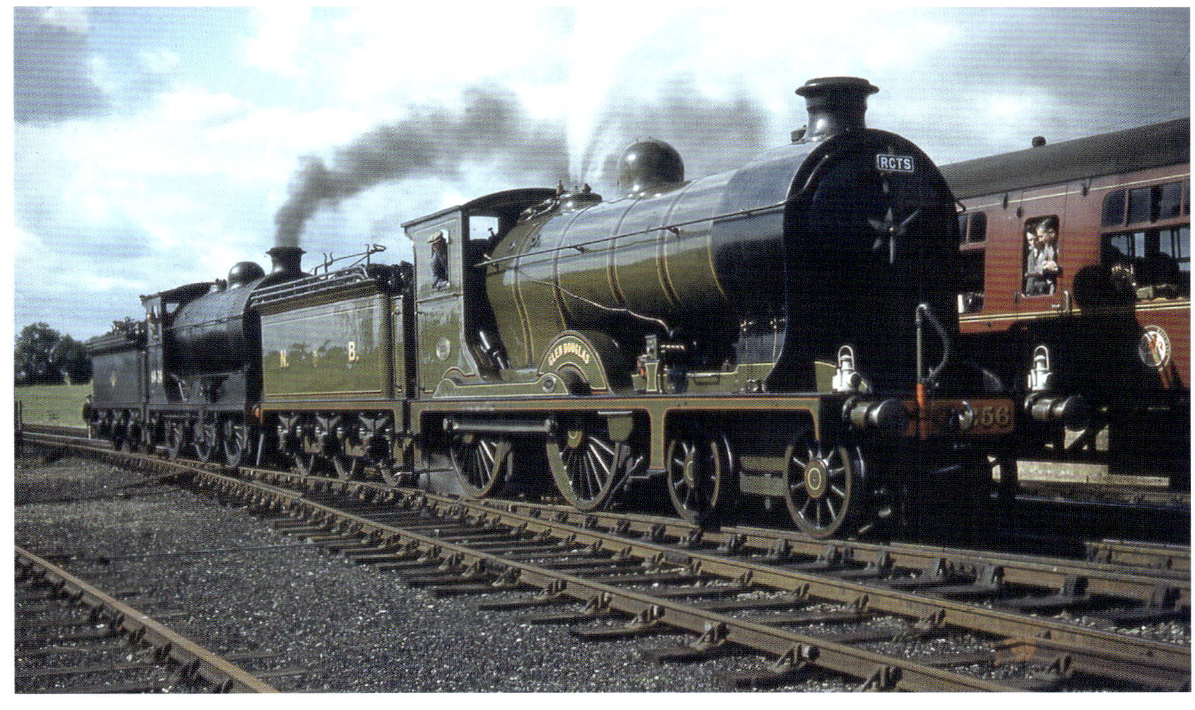

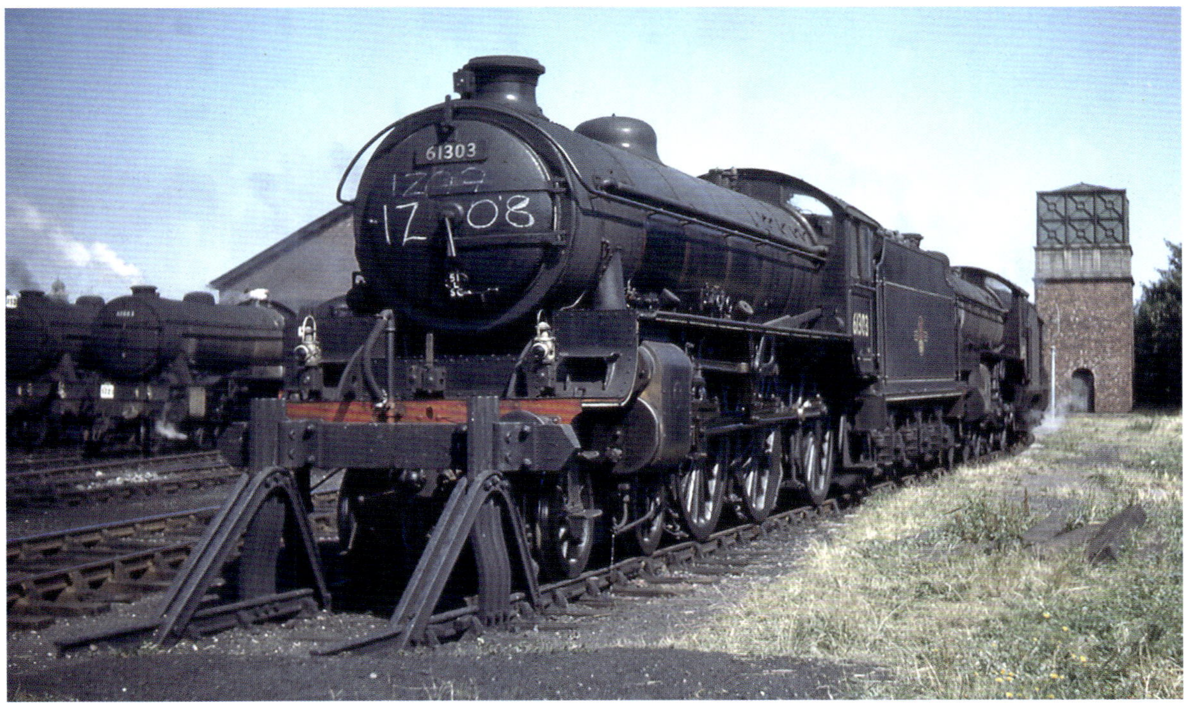

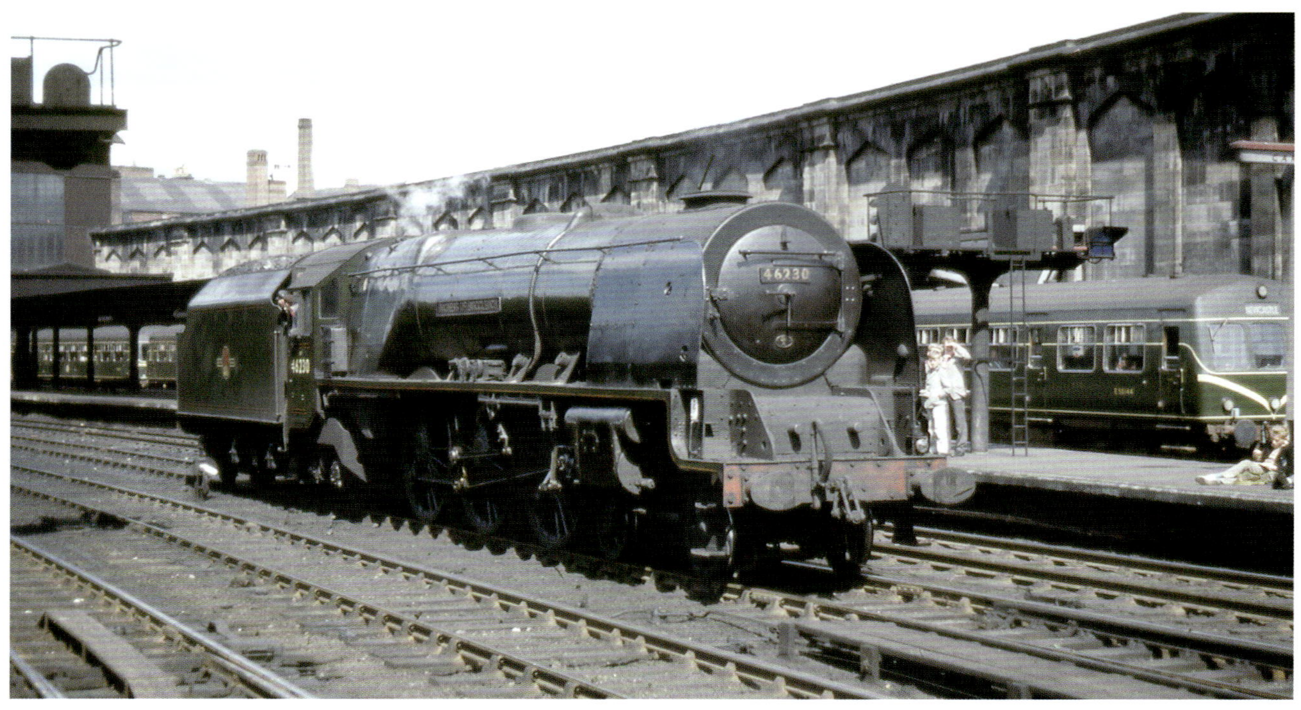

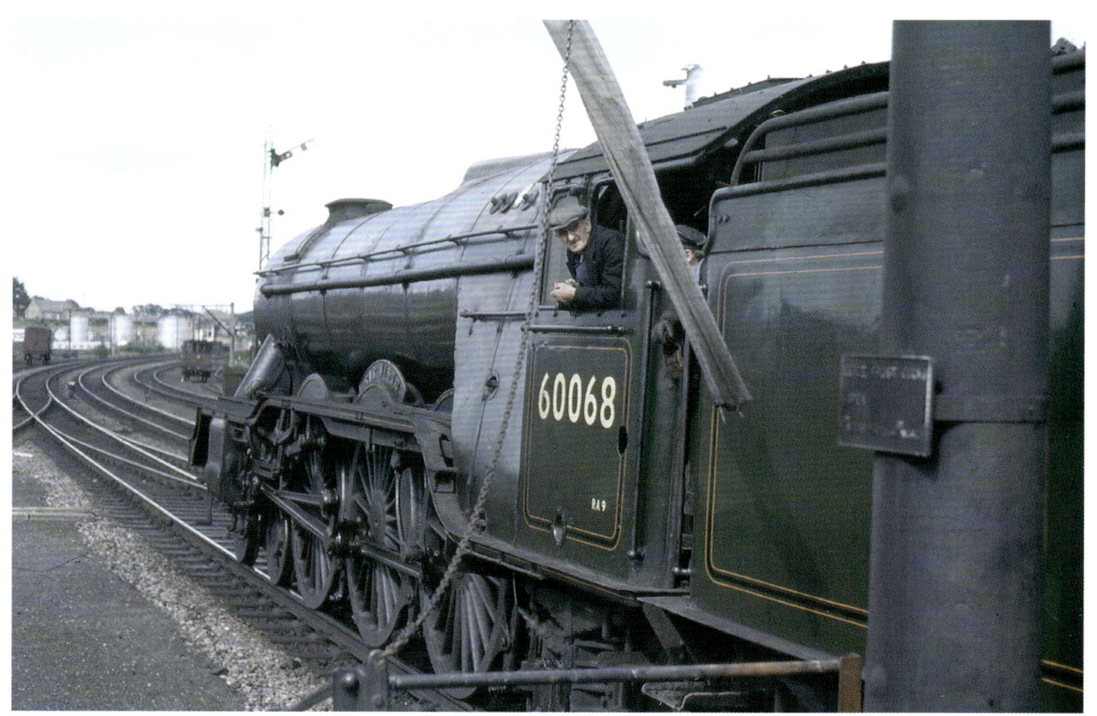

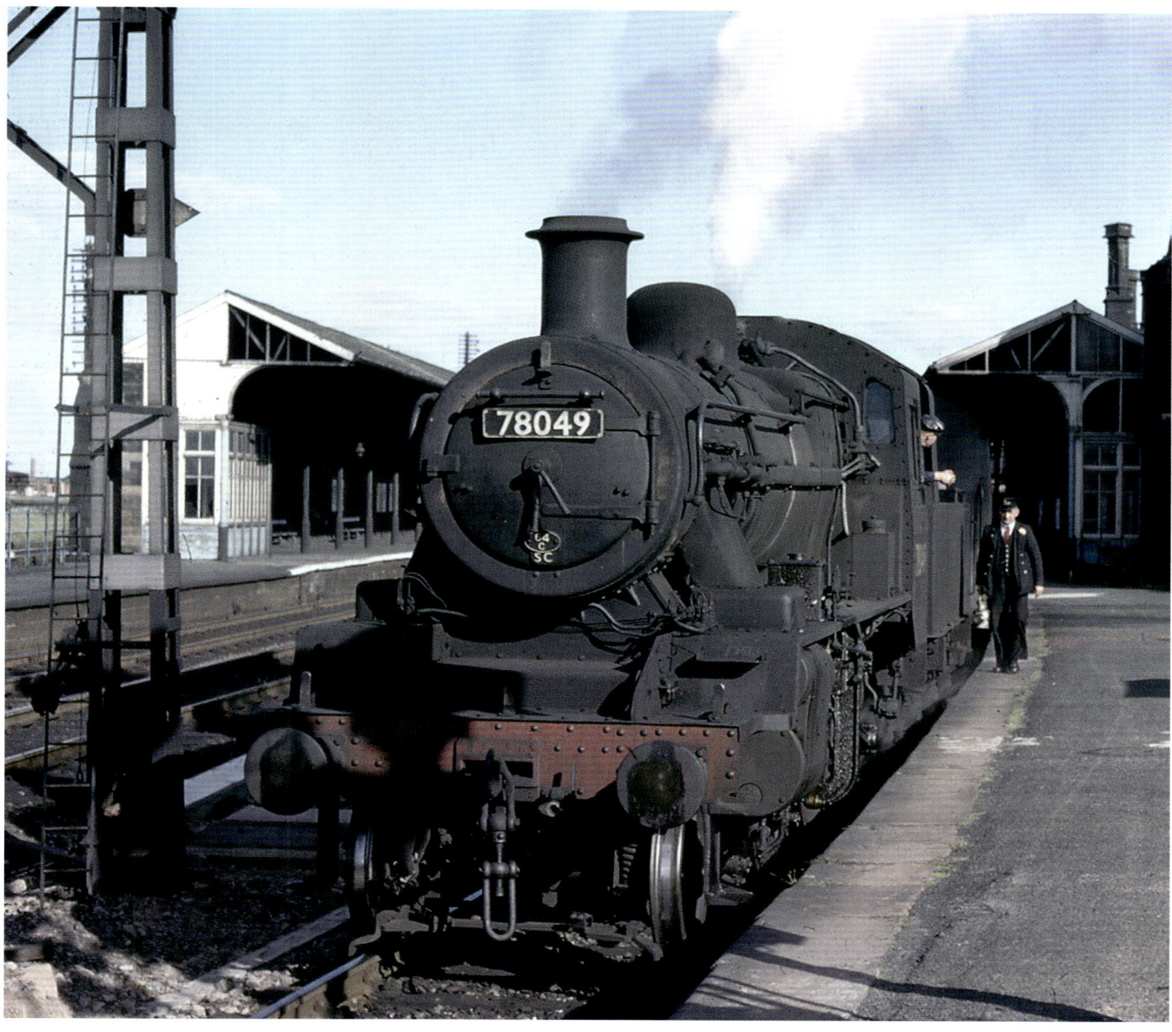

Opposite top: Tuesday 15 August 1961. Waiting for its next duty at Carlisle station is ex-LMS Class 8P Princess Coronation 4-6-2 No. 46230 *Duchess of Buccleuch* sporting green livery. Constructed at Crewe Works during 1938 in the non-streamlined form, she was allocated new to Camden shed and re-allocated to Polmadie during 1944, and is bearing the correct 66A shed code. She would be withdrawn from service in 1963. (Andrew G. Forsyth)

Opposite bottom: Tuesday 15 August 1961. With its driver awaiting 'the right away', ex-LNER Class A3 60068 *Sir Visto* is about to depart from St Boswells station with a 'down' working. Constructed by the NBL in 1924 as a Class A1 locomotive, she was rebuilt as a Class A3 during 1948 and would be withdrawn in 1962. (Andrew G. Forsyth)

Above: Tuesday 15 August 1961. Having arrived at Tweedmouth station from St Boswells via the 'Tweed Valley Line', BR Standard Class 2 2-6-0 No. 78049 has run round its train and is preparing to depart for Berwick-upon-Tweed. Entering service from Darlington Works in 1955, she would be allocated to Scottish sheds for her entire working life, being withdrawn during 1966. The guard is walking forward carrying the lamp for the locomotive. (Andrew G. Forsyth)

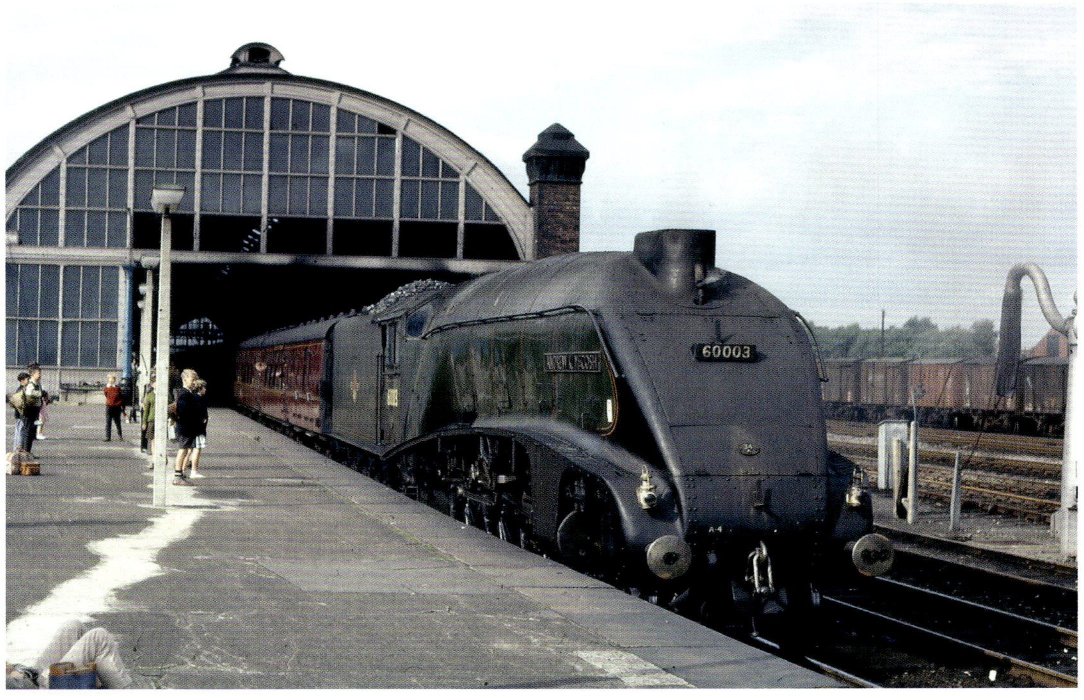

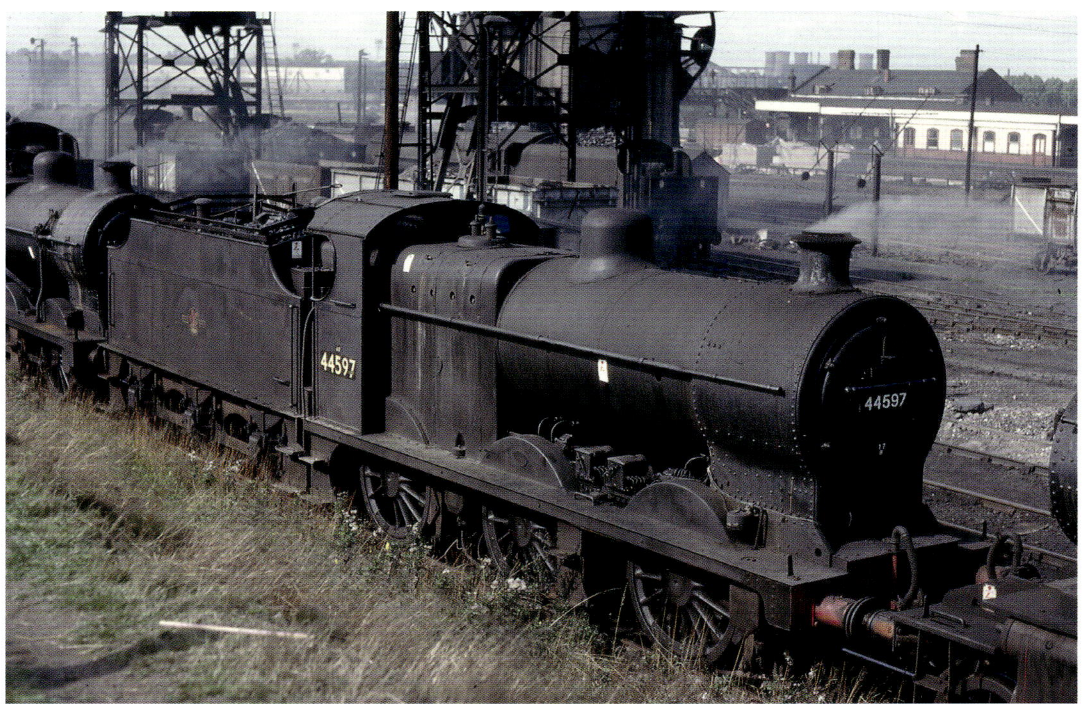

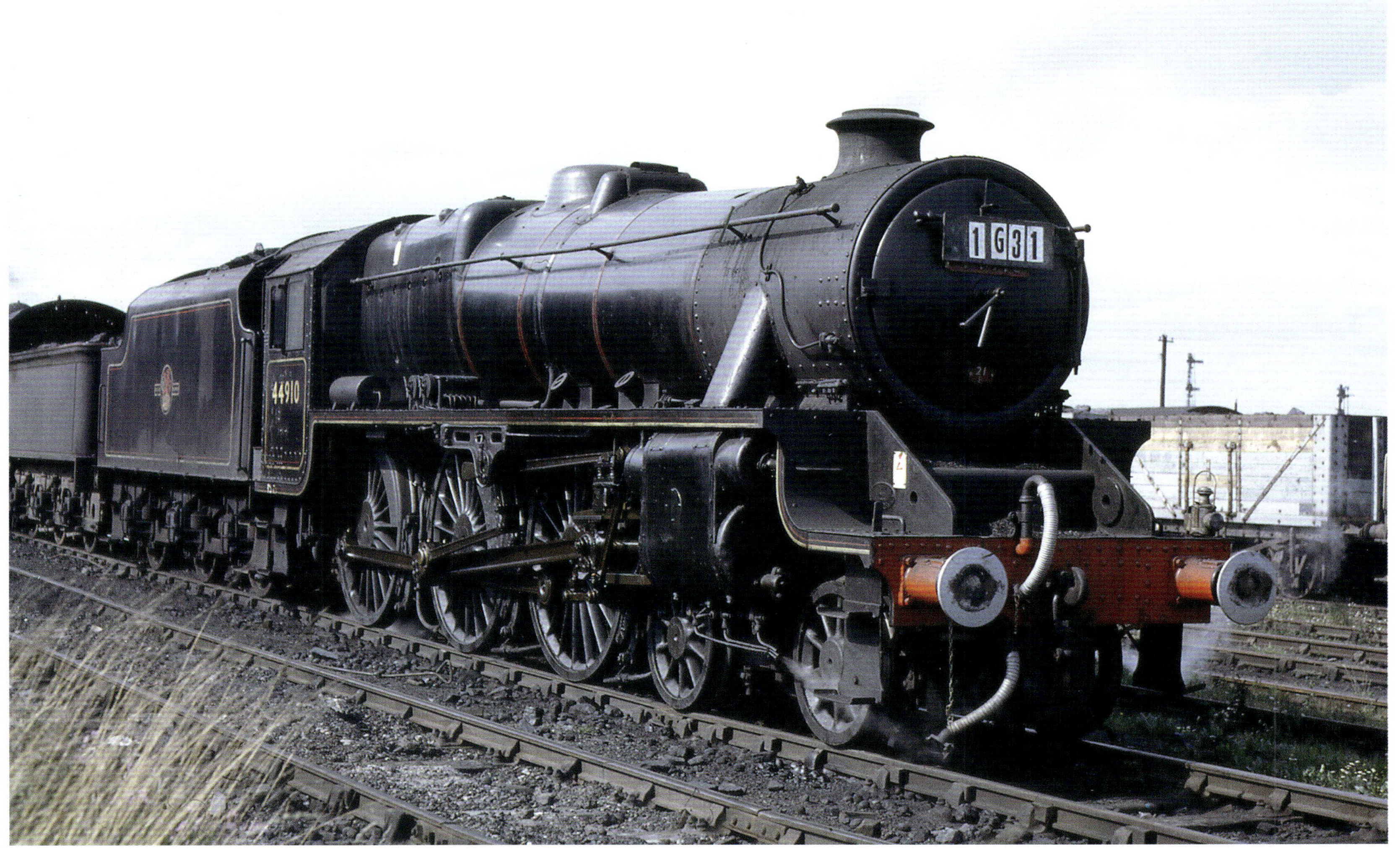

Opposite top: Thursday 17 August 1961. At the head of an 'up' express, ex-LNER Class A4 4-6-2 No. 60003 *Andrew K. McCosh* is waiting to depart from Darlington Bank Top station. Constructed at Doncaster Works during 1937 and originally carrying the name *Osprey*, she was renamed in 1942 to honour the chairman of the LNER Locomotive Committee and would be withdrawn in 1962. (Andrew G. Forsyth)

Opposite bottom: Sunday 27 August 1961. At Bescot shed ex-LMS Class 4F 0-6-0 No. 44597 is bearing a 17B Burton shed code. Constructed at Derby Works during 1940, she would be withdrawn in 1965. (Andrew G. Forsyth)

Above: Sunday 27 August 1961. Also seen at Bescot shed is ex-LMS Class 5 4-6-0 No. 44910, which is bearing a 21E Monument Lane, Birmingham, shed code. Entering service from Crewe Works in 1945, she would be one of the last of her class to be withdrawn during June 1968 whilst allocated to Newton Heath shed. (Andrew G. Forsyth)

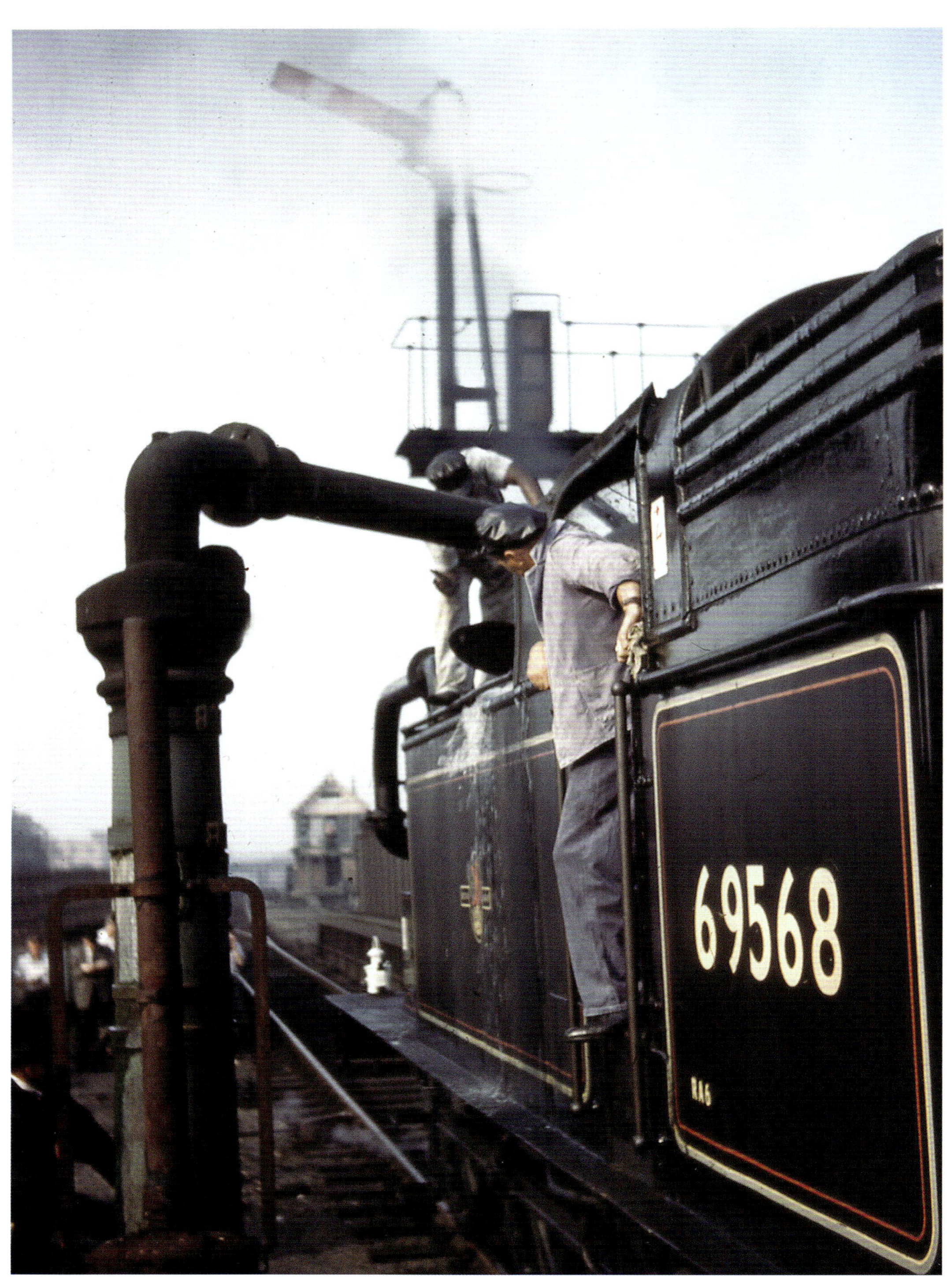

Saturday 2 September 1961. This day saw the running of the Locomotive Club of Great Britain 'North London Rail Tour'. It commenced early in the afternoon at London Marylebone behind ex-LMS Class 3F 2-6-2 tank No. 40031 and proceeding to Moorgate via Neasden, Richmond, Cricklewood, Kentish Town and King Cross Metropolitan. The Moorgate to High Barnet section saw ex-GNR Class N2 (LNER Class N2) 0-6-2 tank No. 69568 in charge to Dalston.

Left: No. 69568 is seen taking water in the 'down slow' platform at Finsbury Park station. (Hugh D. Ramsay)

Opposite top: No. 69568 is seen at the East Finchley stop. The first of the Hawthorn Leslie & Co. constructed members of the class to enter service during 1928, she would be withdrawn in September 1962. (Hugh D. Ramsay)

Opposite bottom: No. 40031 is seen at the former Harlesden station. A product of Derby Works in 1931, she would be withdrawn in December 1962. (Hugh D. Ramsay)

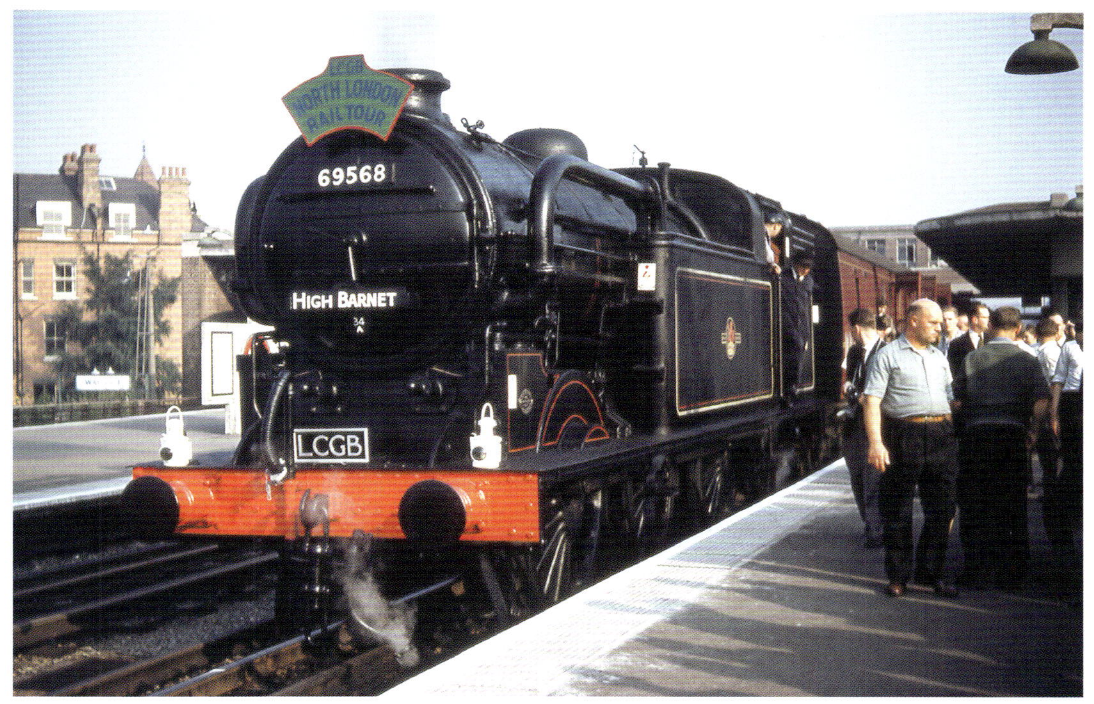
LCGB
NORTH LONDON
RAIL TOUR
69568
HIGH BARNET
34
A
LCGB

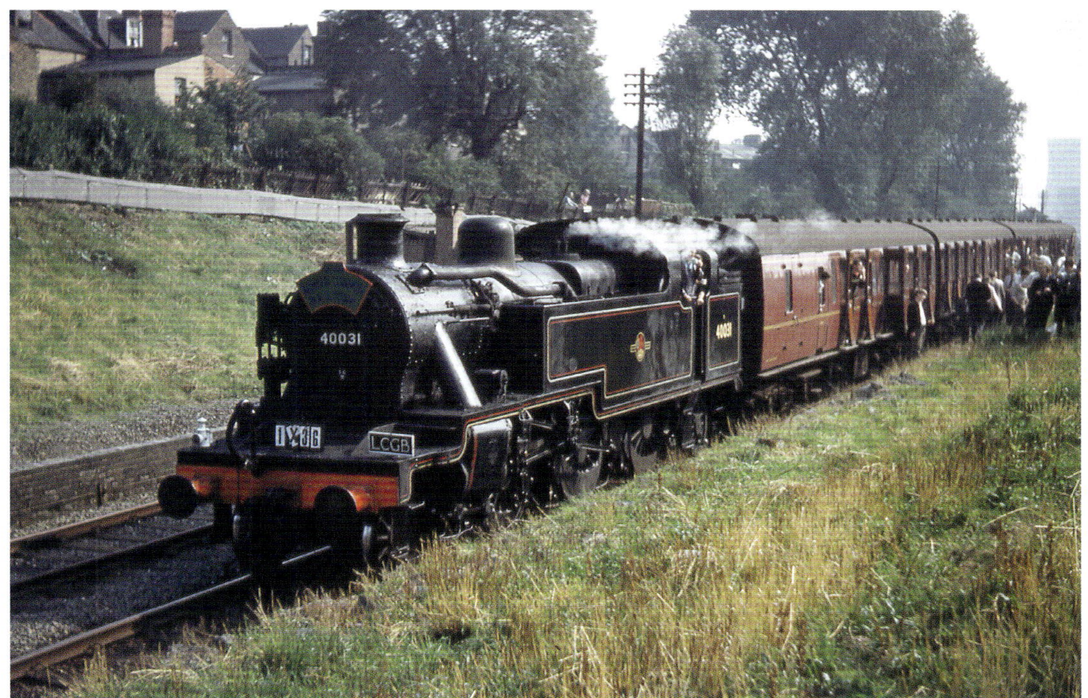
40031
1X86
LCGB

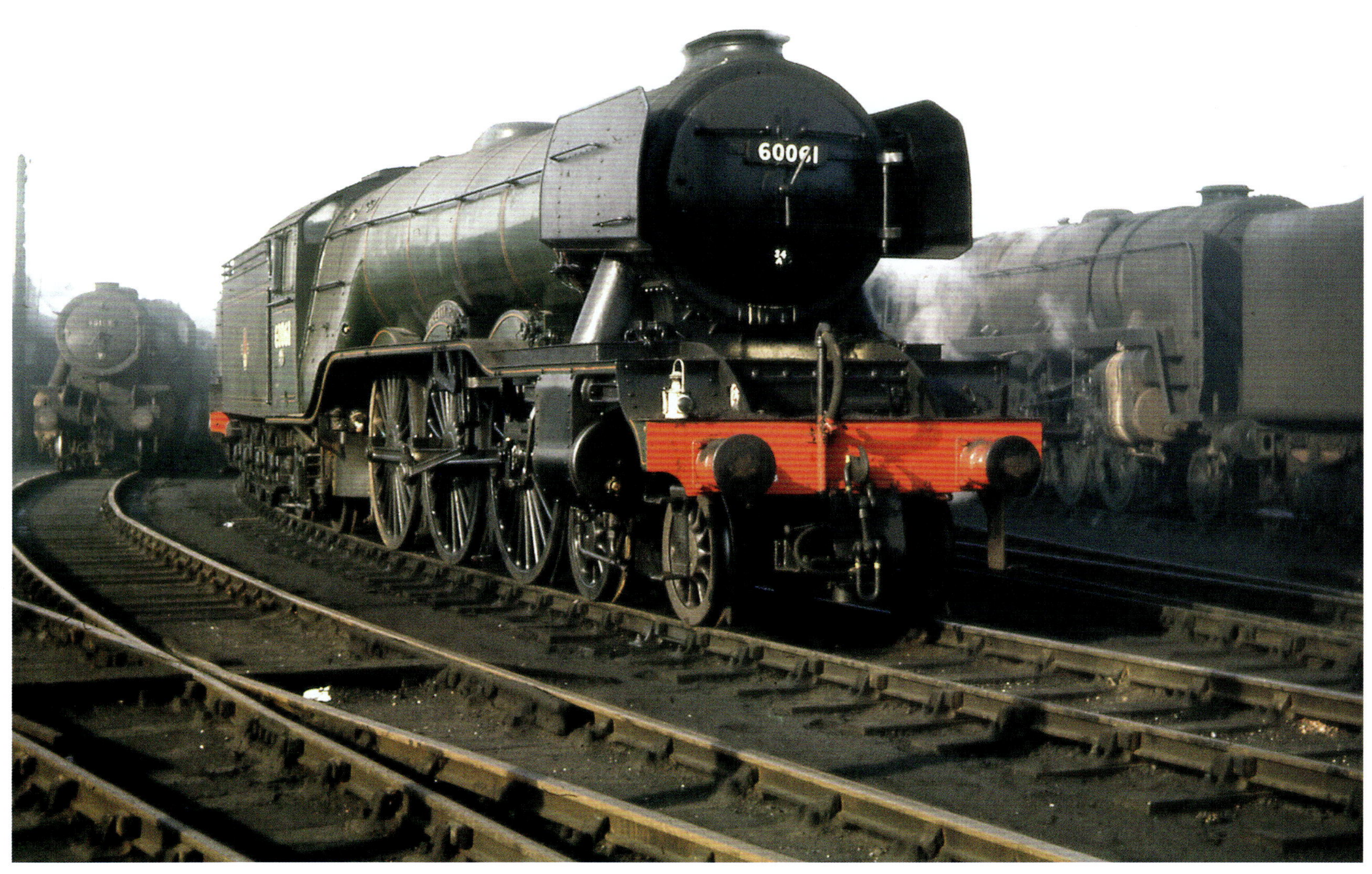

Sunday 11 February 1962. At Kings Cross shed, ex-LNER Class A3 4-6-2 No. 60061 *Pretty Polly* is in beautifully clean condition. Constructed at Doncaster Works during 1925 as a Class A1 locomotive, she would be rebuilt in 1944 as a Class A3, acquiring a double chimney in 1958, trough-type smoke deflectors during 1962 and be withdrawn in 1963. She was named after the racehorse that had won the 1904 Oaks, 1,000 Guineas and the St Leger races. (Hugh D. Ramsay)

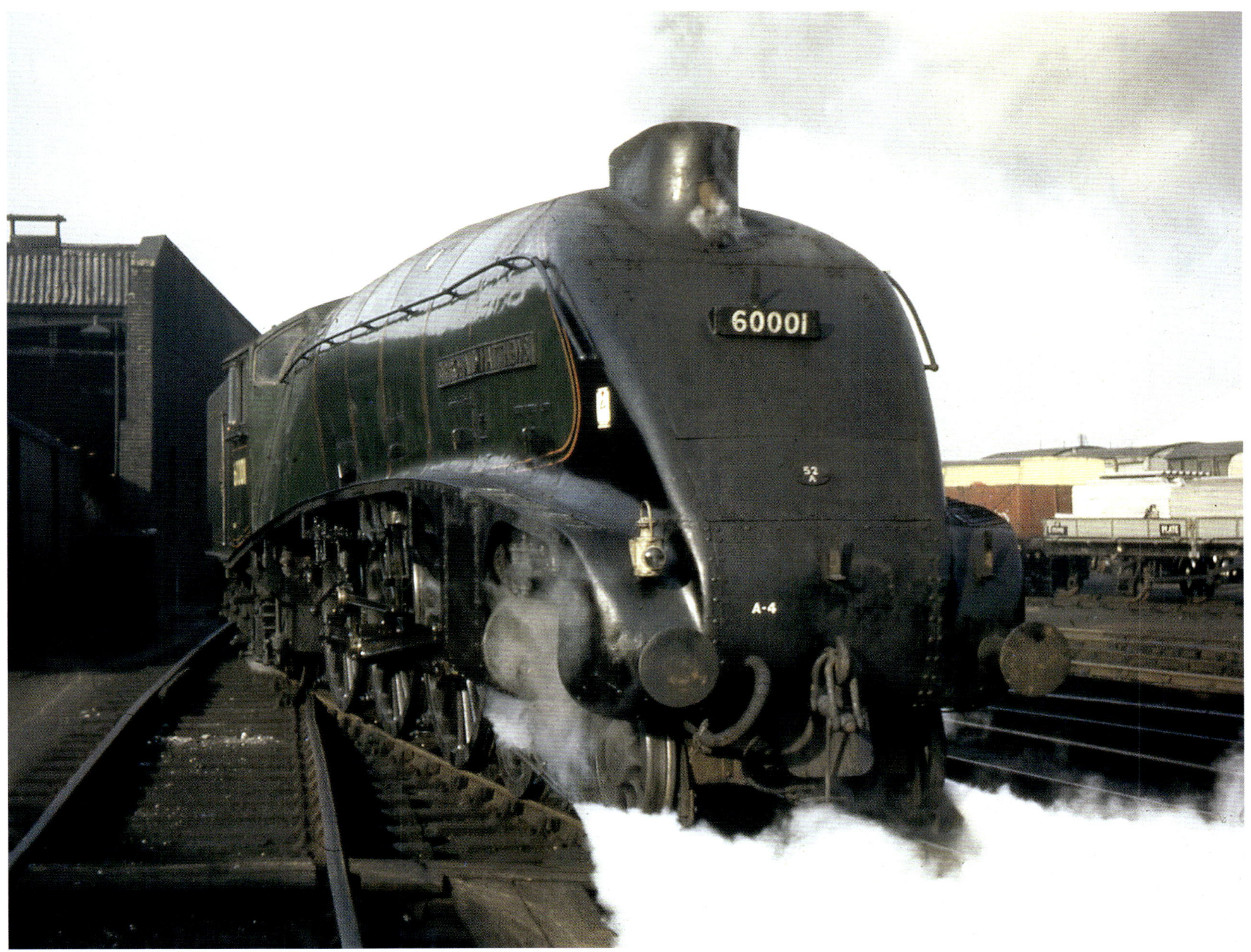

Sunday 11 February 1962. Also at Kings Cross shed and seen manoeuvring in the yard is ex-LNER Class A4 4-6-2 No. 60001 *Sir Ronald Matthews*. Entering service from Doncaster Works during 1938 and originally named *Garganey*, her name was changed in 1939 to honour the chairman of the LNER. Seen bearing a 52A Gateshead shed code, she was allocated there new and would spend her entire working life based there, being withdrawn in 1964. (Hugh D. Ramsay)

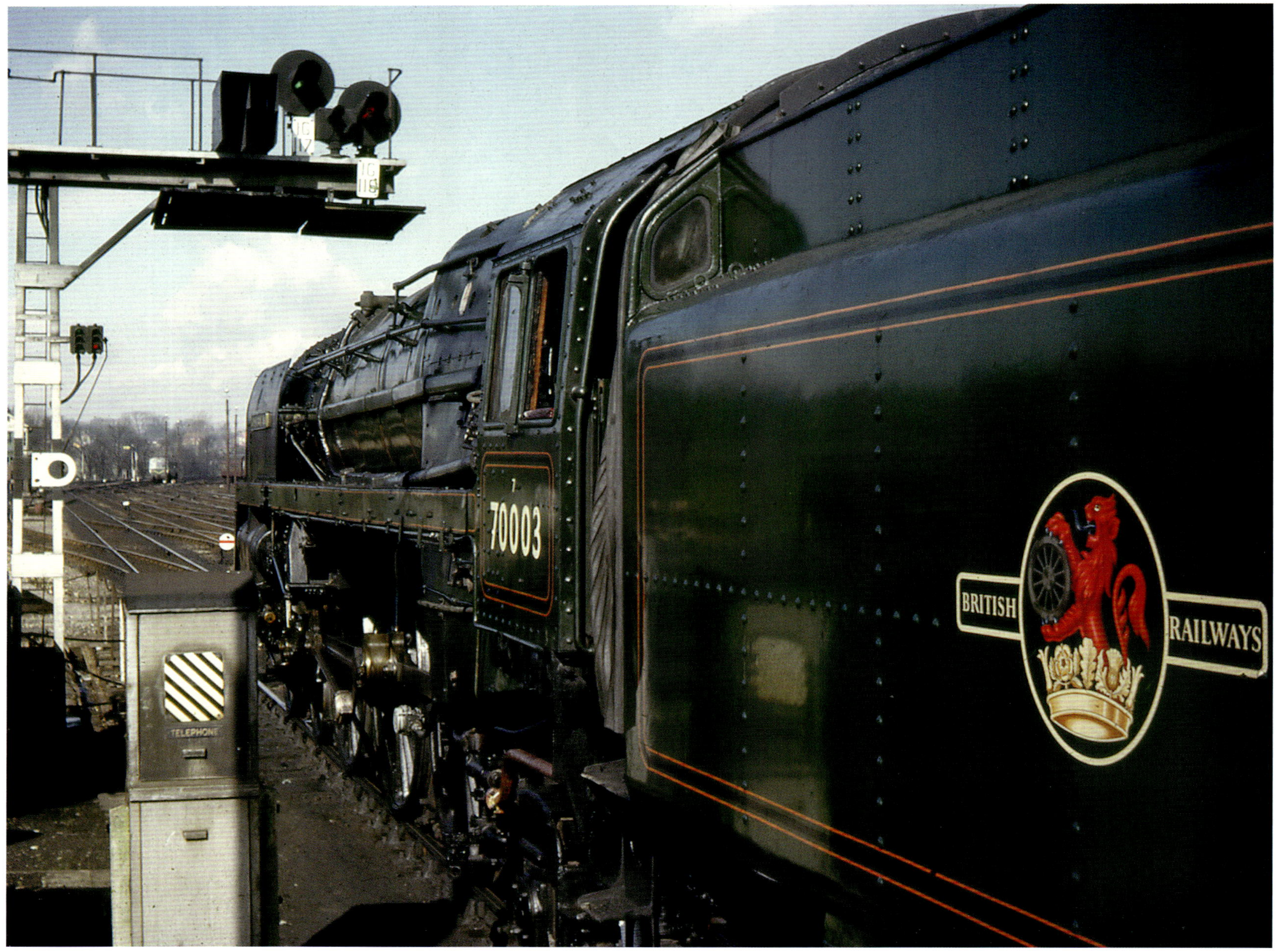

Saturday 31 March 1962. This day saw the running of the RCTS London Branch 'Great Eastern Commemorative Rail Tour', which commenced at London Liverpool Street station, visiting Ipswich, Norwich, Dereham, Swaffham, Thetford, Cambridge and returning to Liverpool Street. Motive power on the day included BR Standard Britannia Class 7P6F 4-6-2 No. 70003 *John Bunyan*, which is seen here waiting to depart from Ipswich station on the 'down' working to Norwich. Entering service from Crewe Works during 1951 and allocated new to Stratford shed, she would end her days based at Kingmoor shed in Carlisle and be withdrawn in 1967. (Hugh D. Ramsay)

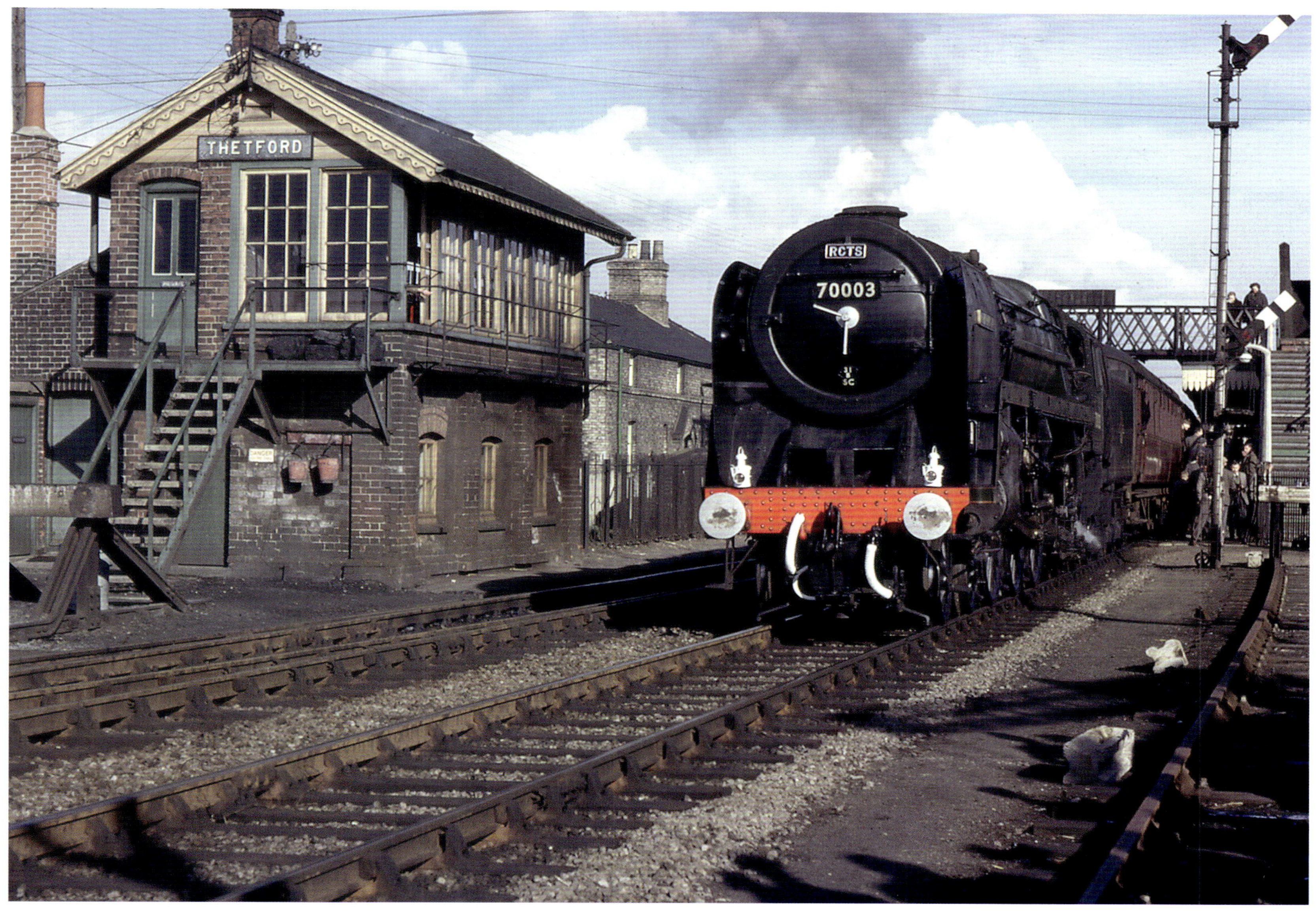

Saturday 31 March 1962. No. 70003 *John Bunyan* is here seen waiting to depart from Thetford station on the return working to Liverpool Street. The Britannia class was the first of the BR Standard classes to be introduced, with No. 70000 *Britannia* herself entering service in January 1951. A total of fifty-five examples were constructed between 1951 and 1954, all coming out of Crewe Works. They were allocated to sheds in the Eastern, Western, London Midland and Scottish regions and many examples survived working until the end of steam traction on British Railways in 1968. (Andrew G. Forsyth)

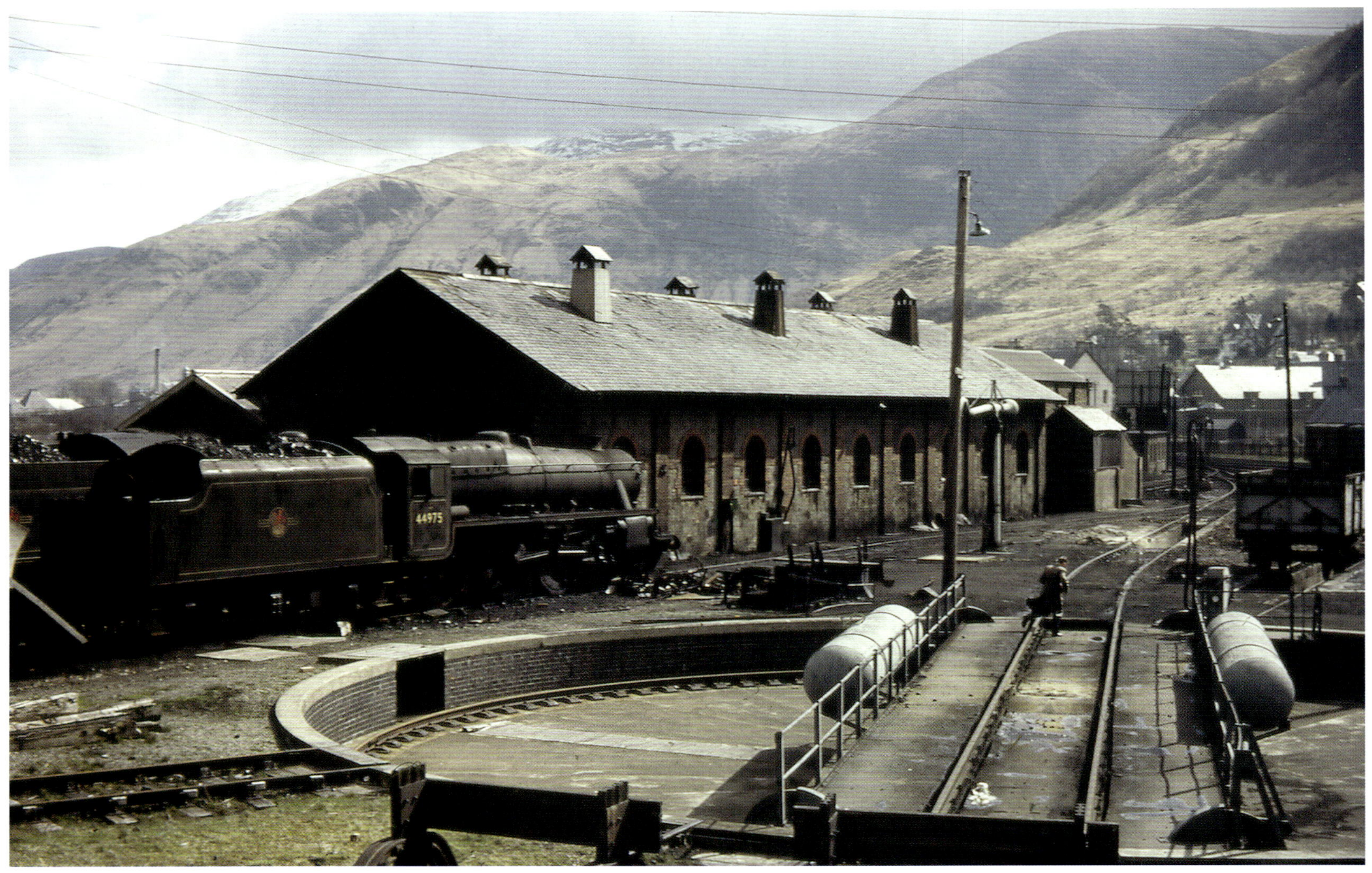

Saturday 21 April 1962. The photographer, on his visit to Fort William, saw the opportunity on this bright day to capture images of ex-LMS Class 5 4-6-0 No. 44975, which is carrying a small snowplough, and ex-LMS Class 4F 0-6-0 No. 44255, which is equipped with a large snowplough.

Above: This general view of the Fort William shed site shows the two-road shed building with the turntable in the foreground. Two months later the shed would be closed to steam traction. (Hugh D. Ramsay)

Opposite top: No. 44975 had entered service from Crewe Works during 1946 and was allocated new to Perth shed. Allocated to Fort William in 1954, she would end her days based at Dalry Road in Edinburgh, being withdrawn during 1965. (Hugh D. Ramsay)

Opposite bottom: No. 44255 had been constructed at Derby Works during 1926 and would be allocated to Fort William shed in 1958. She would be withdrawn from service eight months after this photograph, in December 1962. (Hugh D. Ramsay)

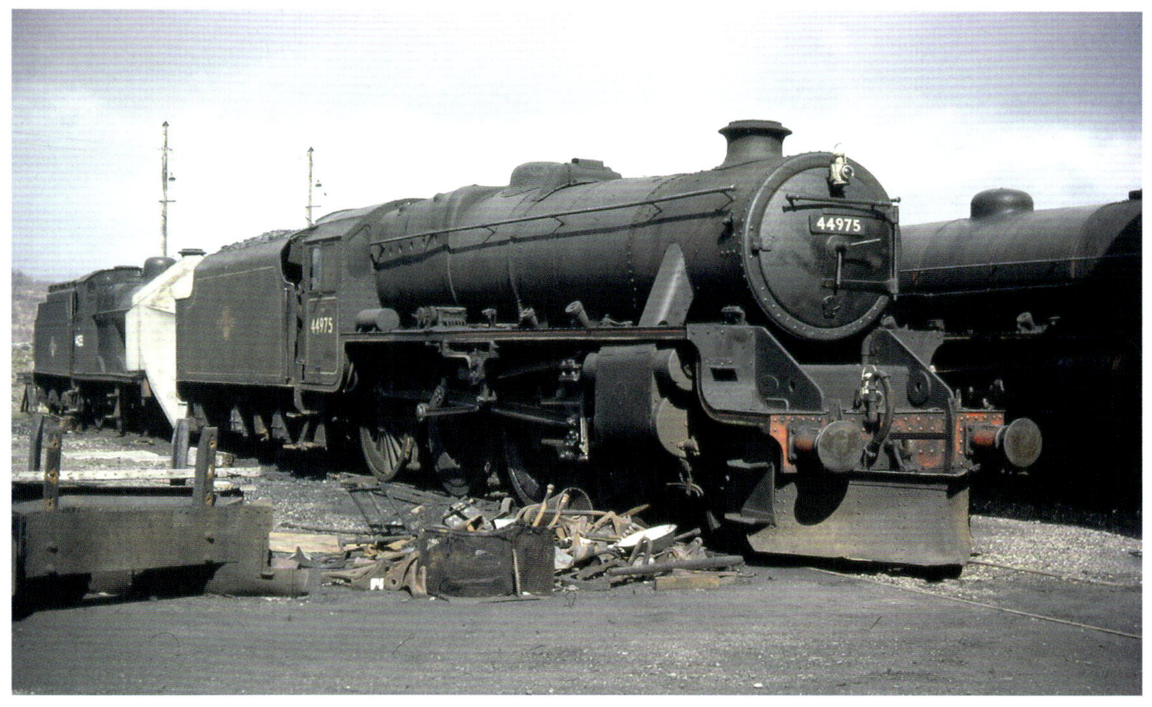

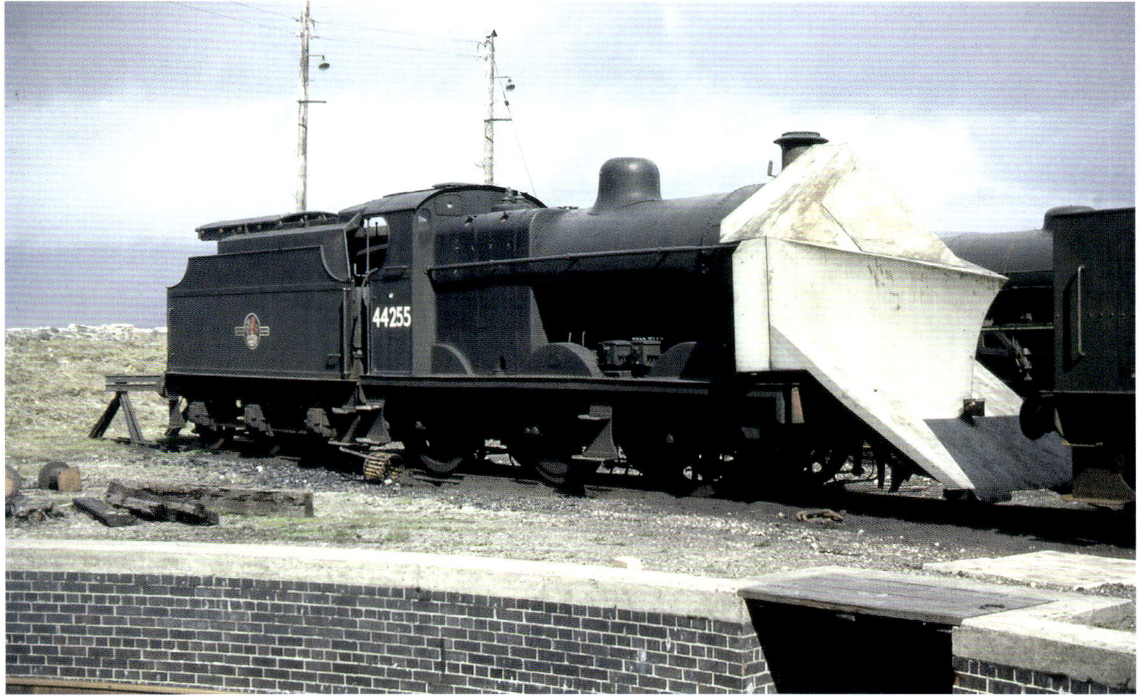

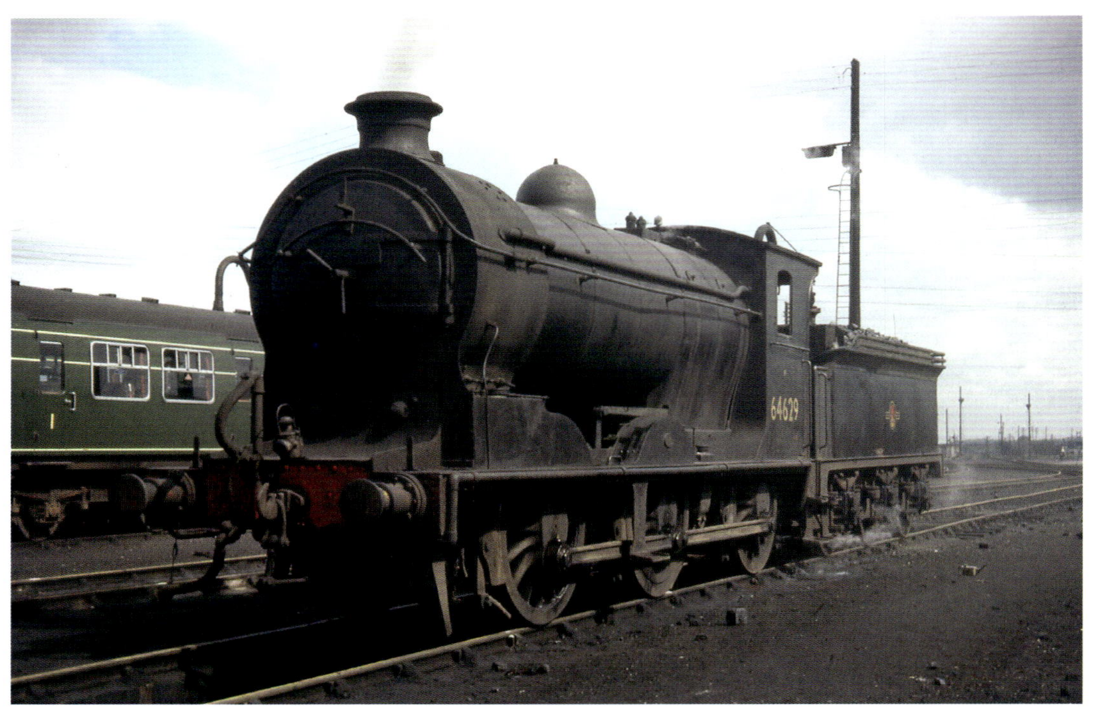

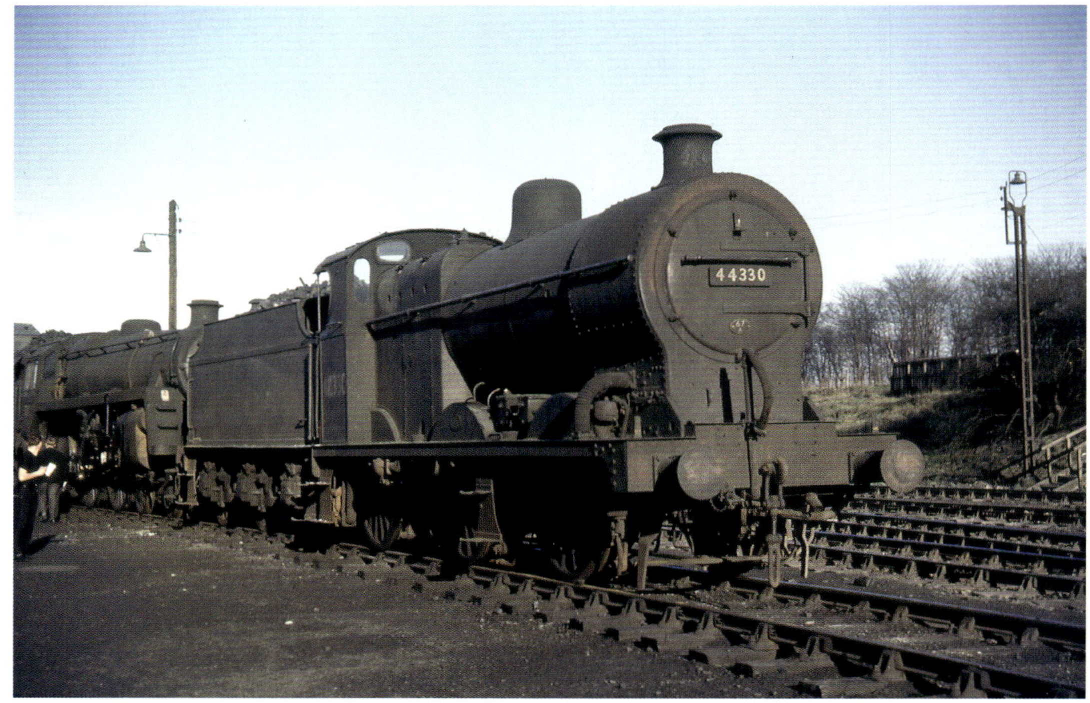

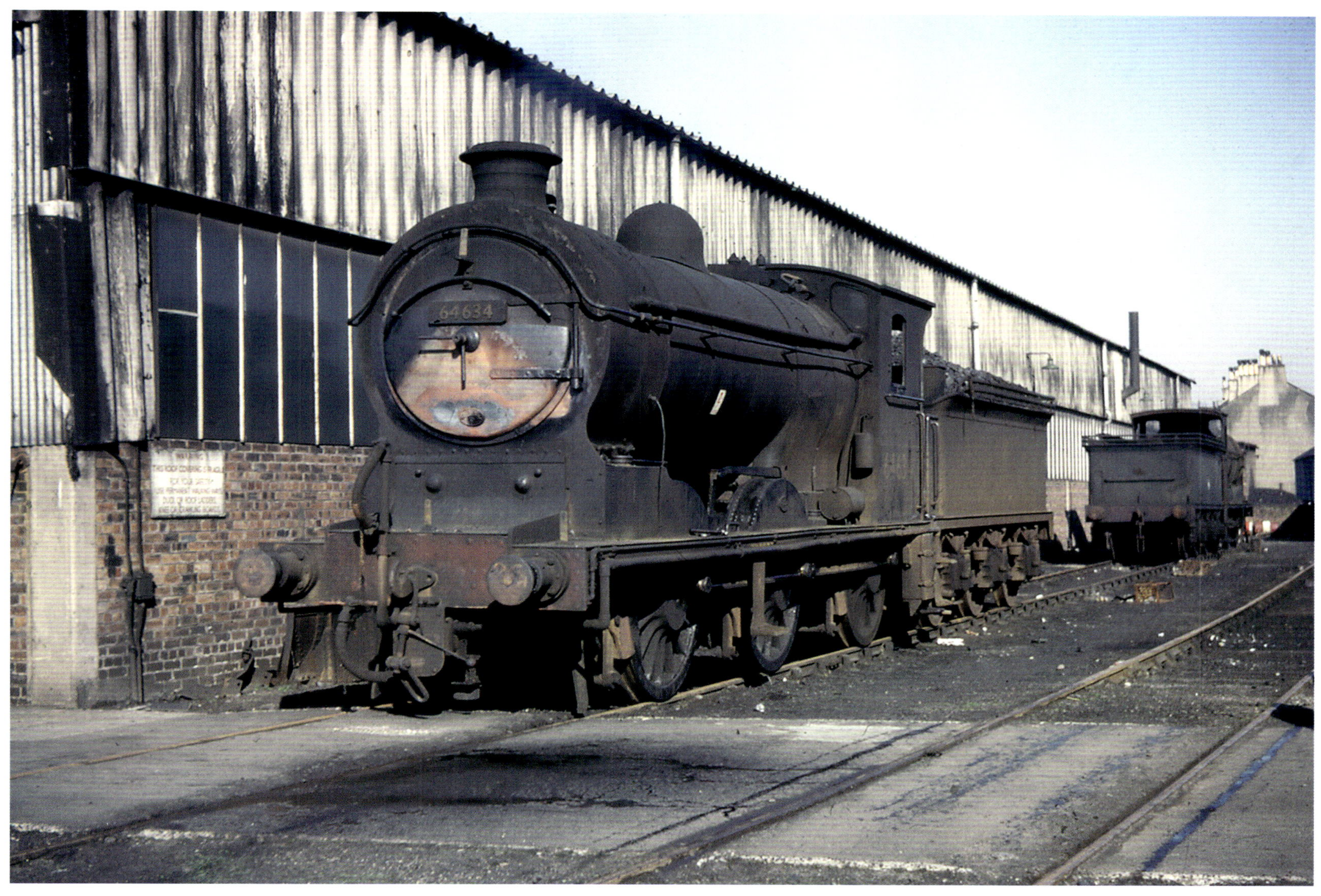

Opposite top: Sunday 22 April 1962. At Thornton Junction shed, ex-NBR Class B (LNER Class J37) 0-6-0 No. 64629 is simmering quietly in the yard. Constructed by the NBL during 1921, she would be withdrawn in 1963. (Hugh D. Ramsay)

Opposite bottom: Sunday 22 April 1962. Ex-LMS Class 4F 0-6-0 No. 44330 is seen in the yard at Corkerhill shed in Glasgow. Entering service from St Rollox Works during 1928, she was allocated to Scottish sheds throughout her working life, being withdrawn later in 1962. (Hugh D. Ramsay)

Above: Monday 23 April 1962. At Bathgate shed, ex-NBR Class B (LNER Class J37) 0-6-0 No. 64634 has suffered some serious damage to the smoke box door. The product of Cowlairs Works during 1921, she would be withdrawn in 1964. (Hugh D. Ramsay)

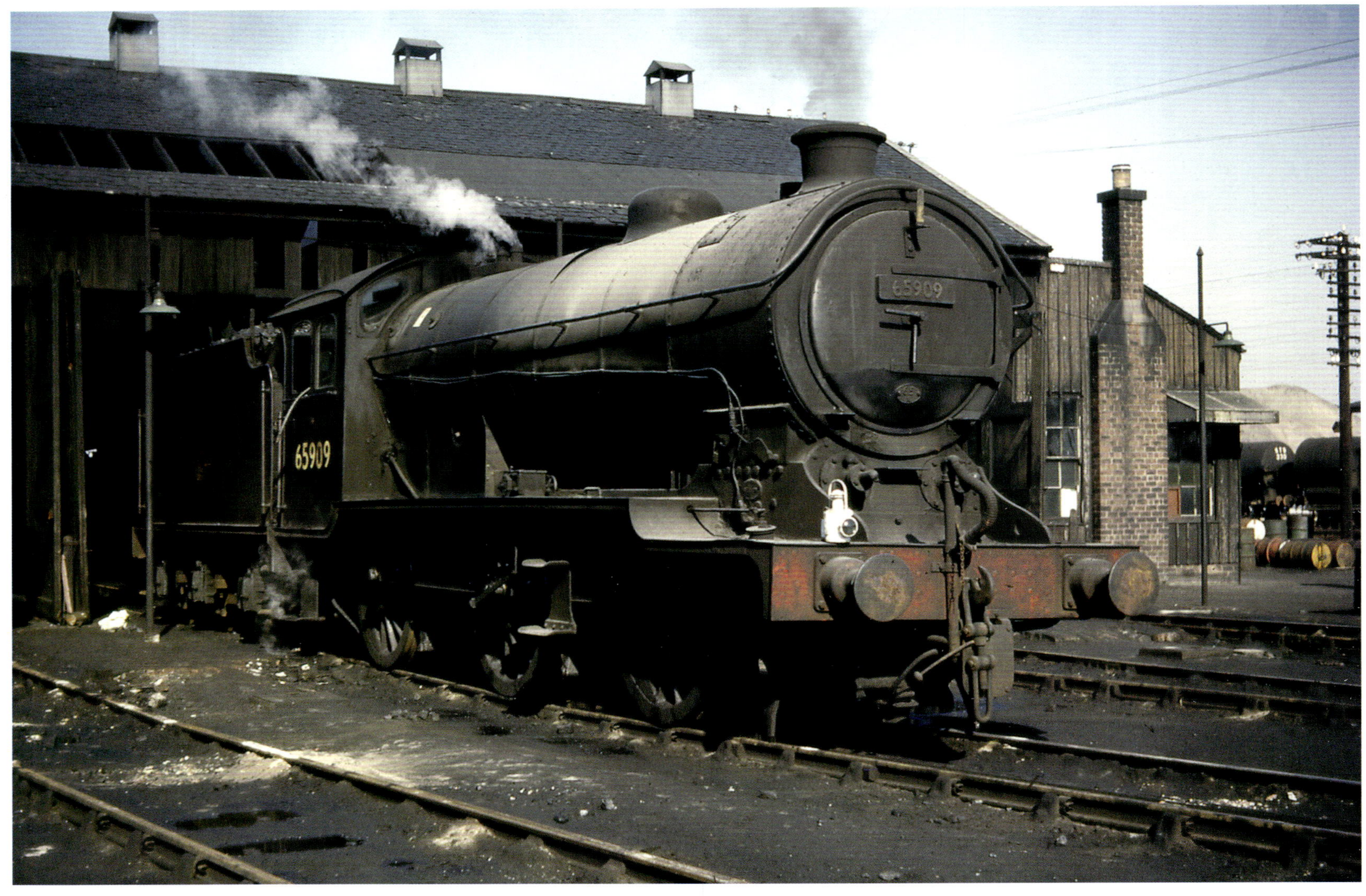

Above: Monday 23 April 1962. Standing adjacent to Polmont shed is ex-LNER Class J38 0-6-0 No. 65909. Constructed as one of a batch of thirty-five members of the class entering service from Darlington Works during 1926, she would give forty years of service before being withdrawn in 1966. (Hugh D. Ramsay)

Opposite top: Monday 23 April 1962. At Grangemouth shed, ex-NBR Class C (LNER Class J36) 0-6-0 No. 65257 had been constructed by Sharp Stewart & Co. in 1892 and would be rebuilt in the form seen here during 1919. She would be withdrawn later during 1962. (Hugh D. Ramsay)

Opposite bottom: Monday 23 April 1962. Parked up at Haymarket shed and waiting a move to Doncaster Works is ex-LNER Class A4 4-6-2 No. 60012 *Commonwealth of Australia*. Entering service from the same works during 1937 and allocated new to Haymarket, she would end her days working out of Ferryhill in Aberdeen, being withdrawn during 1964. (Hugh D. Ramsay)

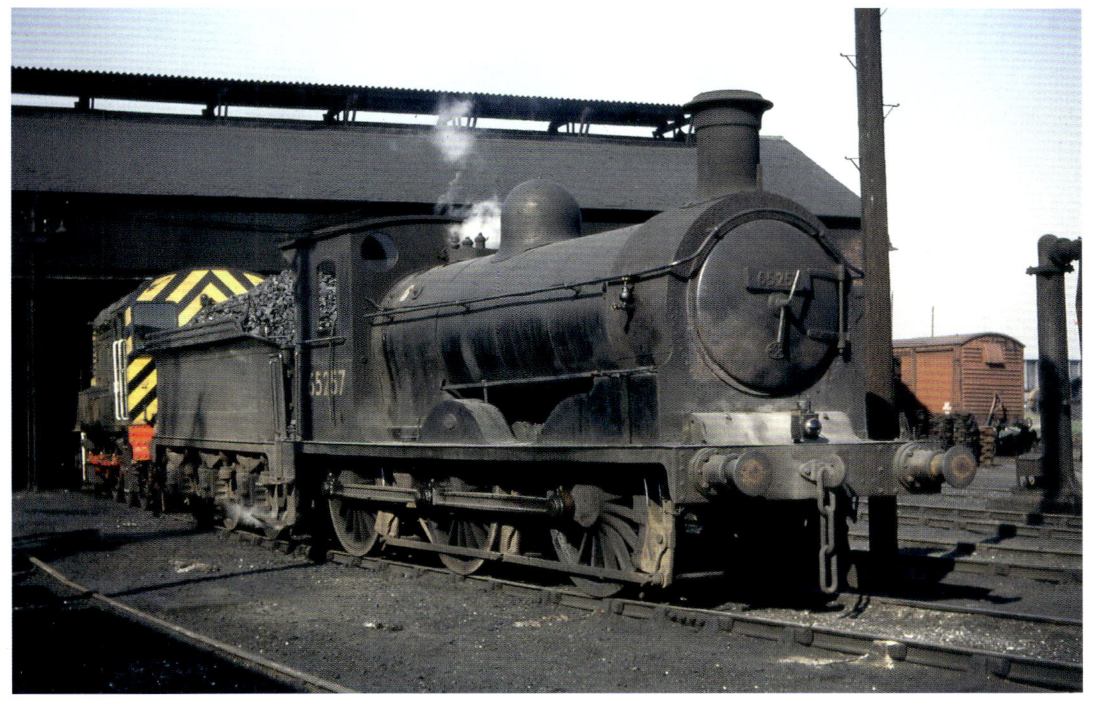

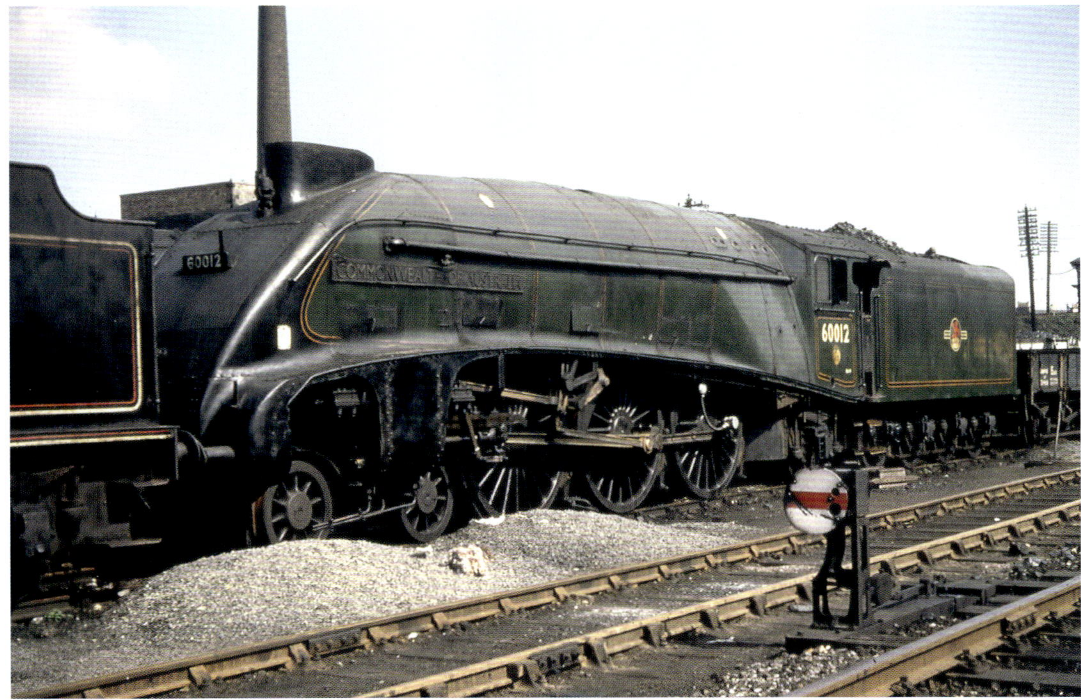

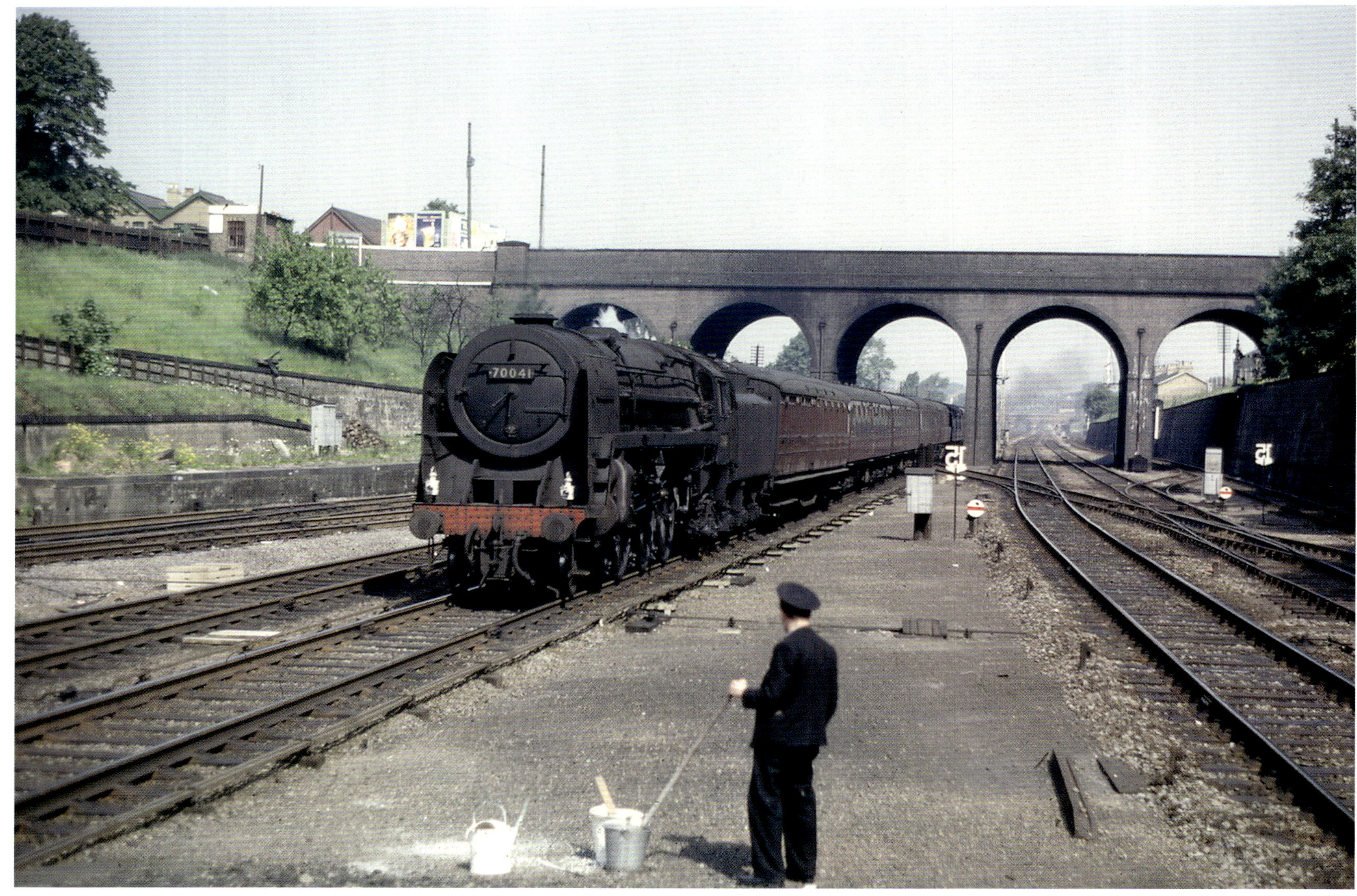

May 1962. The station porter at New Southgate station pauses from mixing platform edge whitening as BR Standard Class 7P6F Britannia 4-6-2 No. 70041 *Sir John Moore* passes at the head of an 'up' express. Constructed at Crewe Works during 1953, she would only give fourteen years of service, being withdrawn in 1967. (Andrew G. Forsyth)

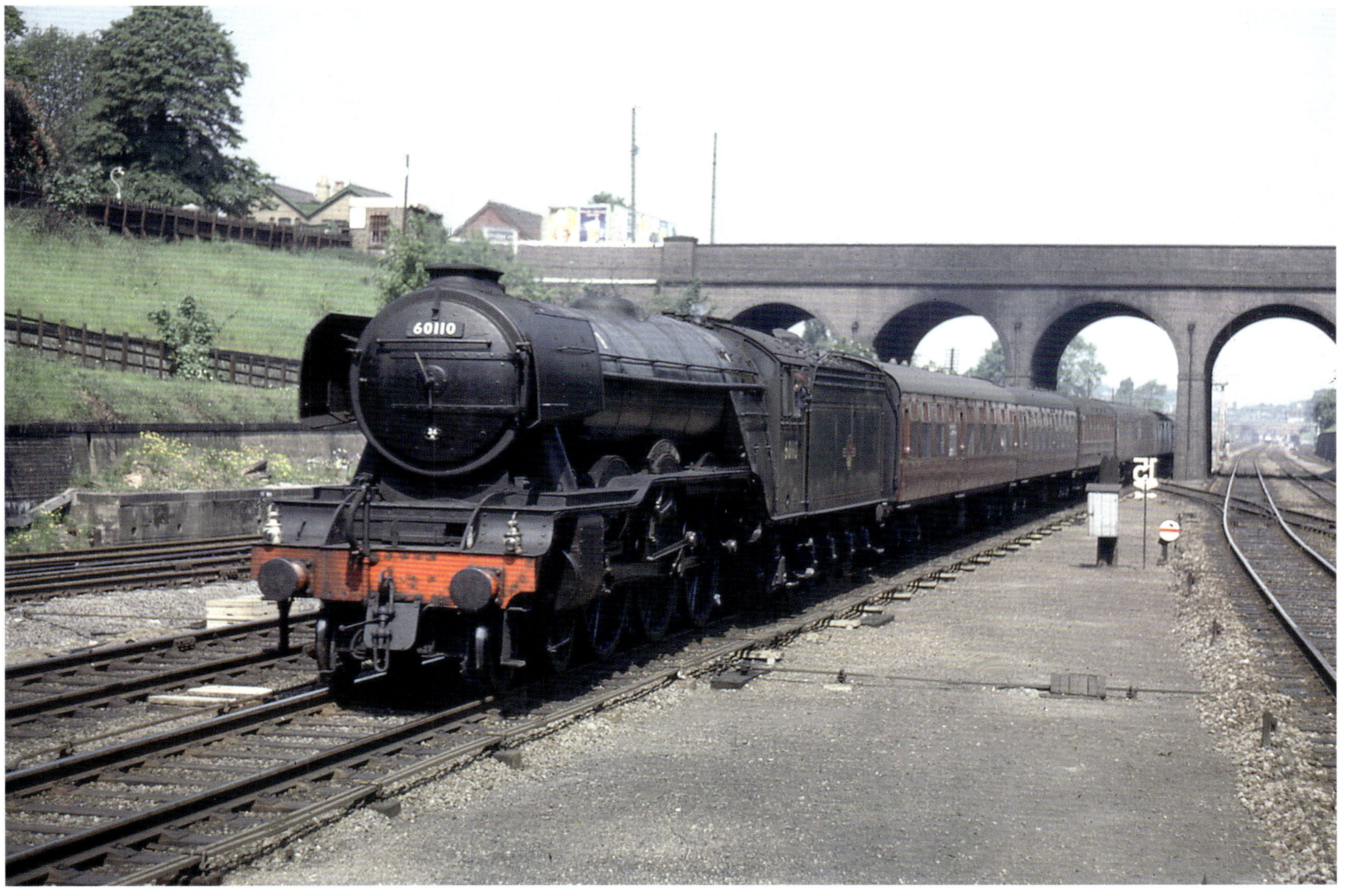

May 1962. The porter at New Southgate station has completed his task so misses ex-LNER Class A3 4-6-2 No. 60110 *Robert the Devil* passing at the head of another 'up' express. Entering service from Doncaster Works during 1923 as a Class A1 locomotive, she would be rebuilt as a Class A3 locomotive in 1942. Named after the racehorse that won the 1880 St Leger race, she would be withdrawn in 1963. (Andrew G. Forsyth)

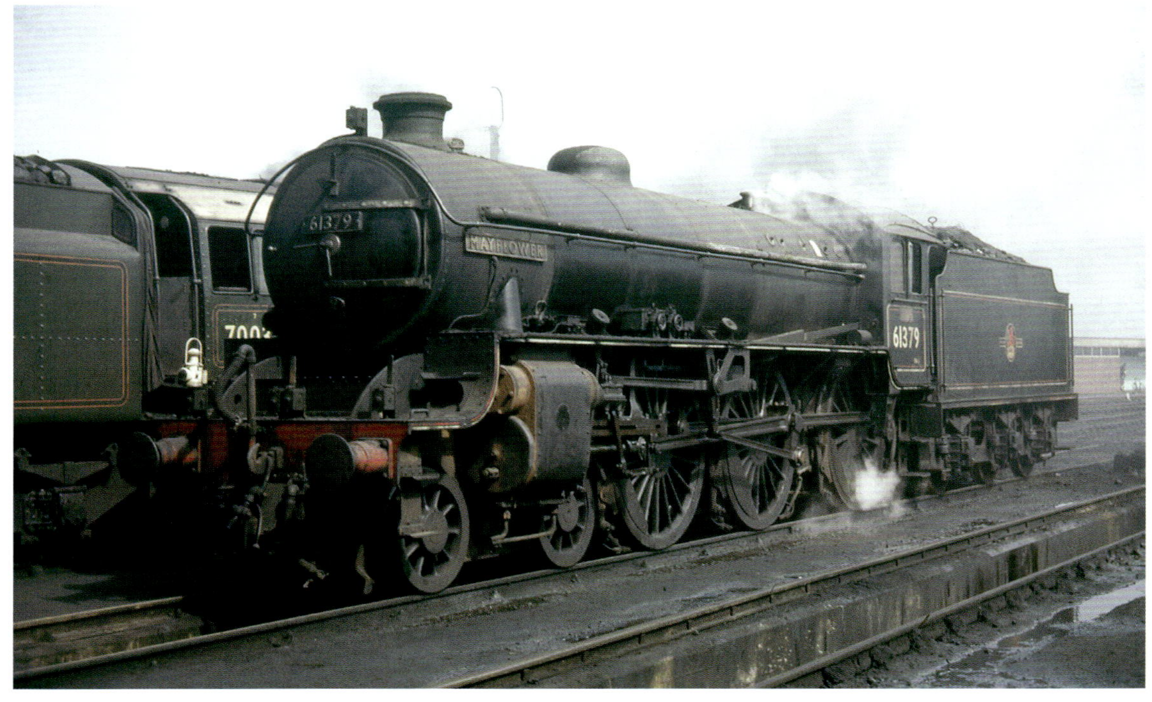

Saturday 12 May 1962. At Immingham shed, Class B1 4-6-0 No. 61379 *Mayflower* is carrying the correct 40B shed code. Constructed by the NBL during 1951, she would be one of the early class withdrawals a mere three months after this photograph was taken, in August 1962. (Hugh D. Ramsay)

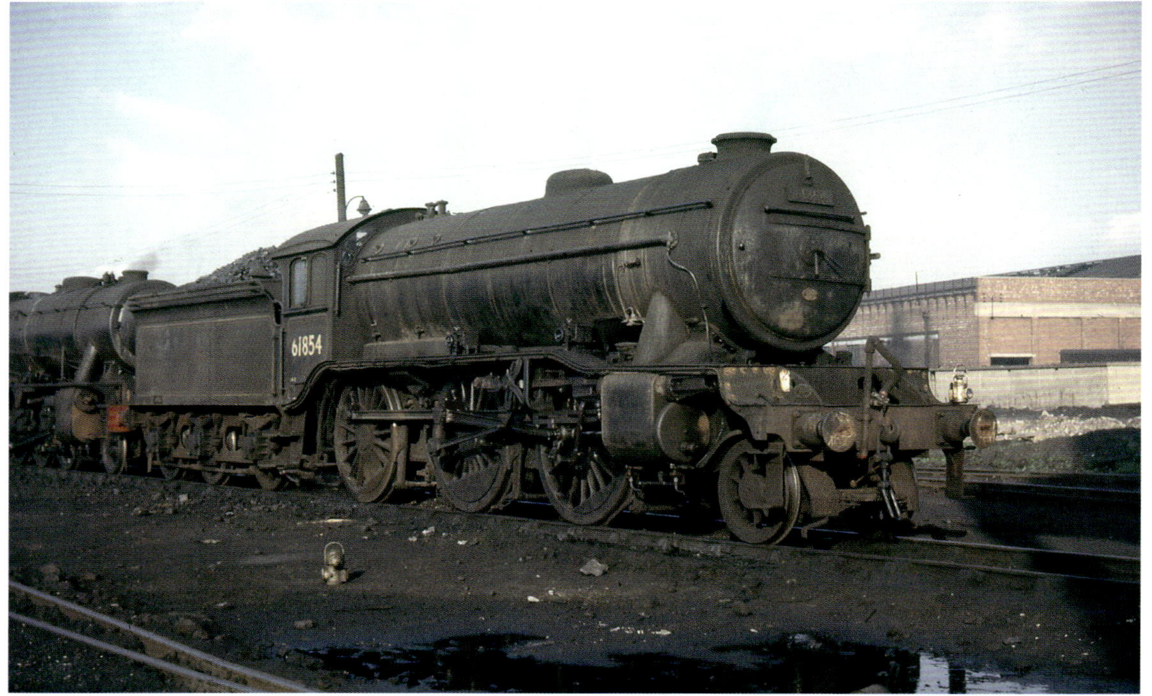

Saturday 12 May 1962. Ex-LNER Class K3 2-6-0 No. 61854 is seen in filthy condition when photographed at Hull Dairycoates shed. Constructed at Darlington Works during 1925, she would be withdrawn from service five months after this photograph was taken. (Hugh D. Ramsay)

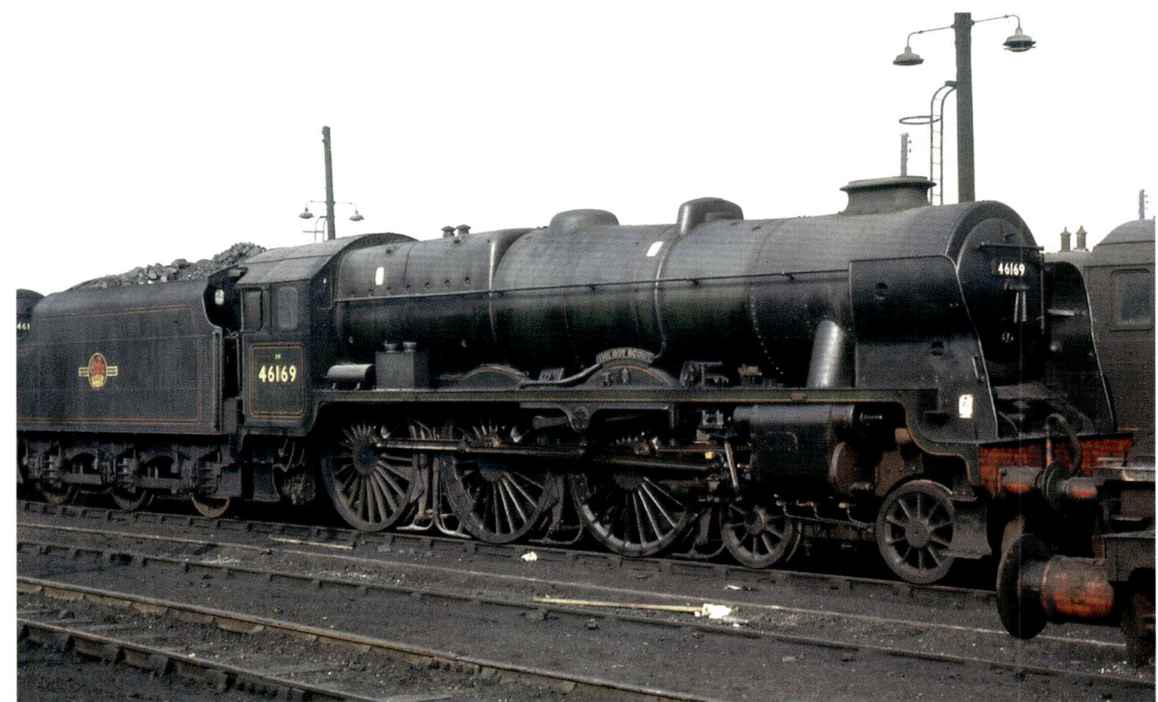

Sunday 20 May 1962. In typical condition for goods locomotives of the period, ex-LMS Class 5 2-6-0 No. 42833 is seen at Lancaster Green Ayre shed. Seen here carrying the correct 24J shed code and fitted with a small snowplough, she had been the product of Horwich Works in 1930 but would only give thirty-two years of service before being withdrawn during December 1962. (Hugh D. Ramsay)

June 1962. At Willesden shed, a reasonably clean ex-LMS Class 7P Royal Scot 4-6-0 No. 46169 *The Boy Scout* is standing in a row of parked locomotives. Originally constructed at Derby Works in parallel boiler form during 1930, she would be rebuilt with a taper boiler in 1945 and withdrawn from service during 1963. (Anthony Scarsbrook)

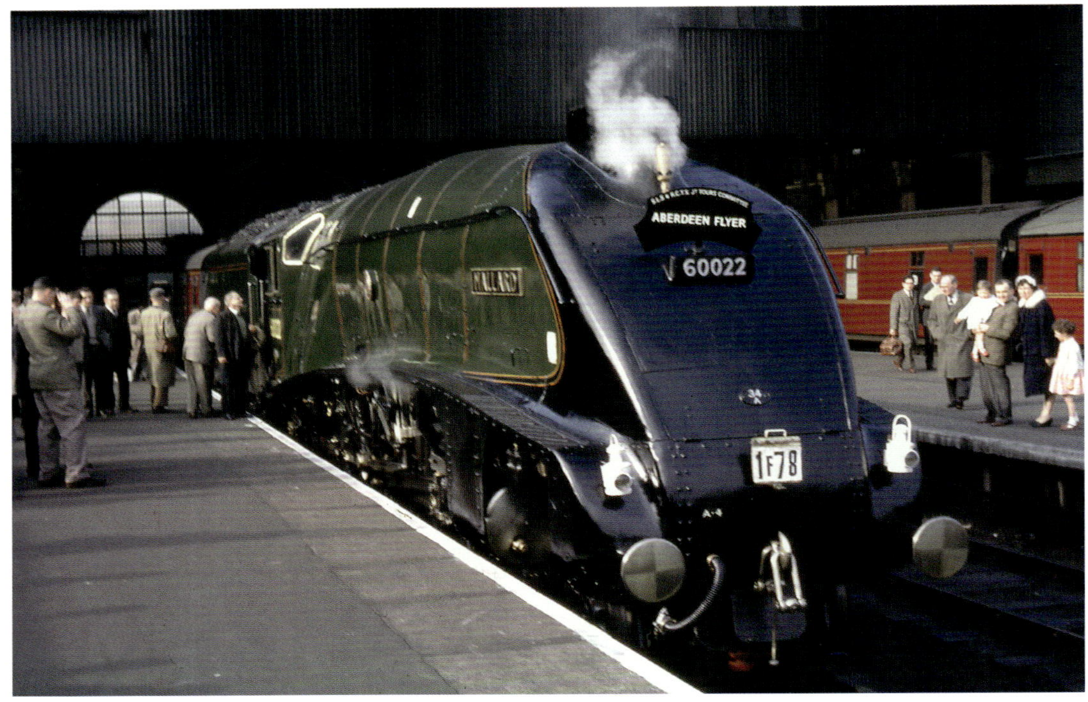

Saturday 2 June 1962. This day saw the running of the joint RCTS and SLS 'Aberdeen Flyer Rail Tour', commencing at London Kings Cross, with the route taking it through Edinburgh Waverley to reach Aberdeen. Returning from Aberdeen to Carlisle overnight, the tour would end at London Euston. Locomotives utilised included ex-LNER Class A4 4-6-2 No. 60022 *Mallard*, working the train from Kings Cross to Edinburgh Waverley, and ex-LMS Class 8P Princess Royal 4-6-2 No. 46200 *The Princess Royal*, working from Carlisle to Euston.

Left: No. 60022 *Mallard* is seen in immaculate condition at Kings Cross prior to the departure of the rail tour. (Anthony Scarsbrook)

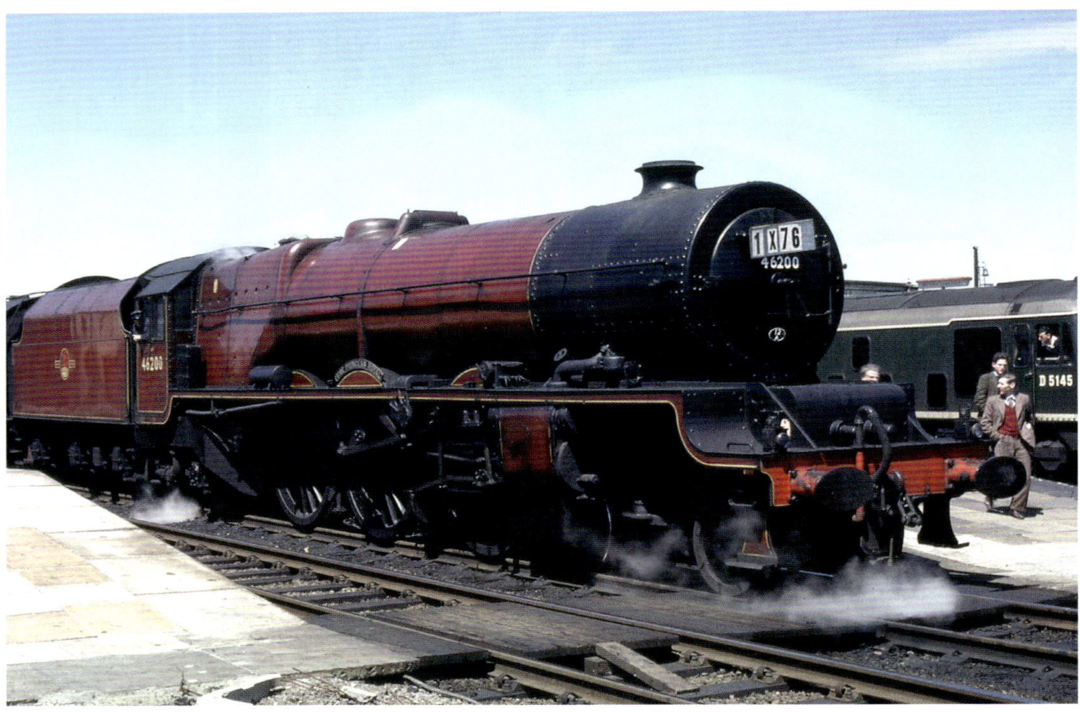

Sunday 3 June 1962. At Watford station, No. 46200 *The Princess Royal* has paused to allow passengers to alight from the train. (Anthony Scarsbrook)

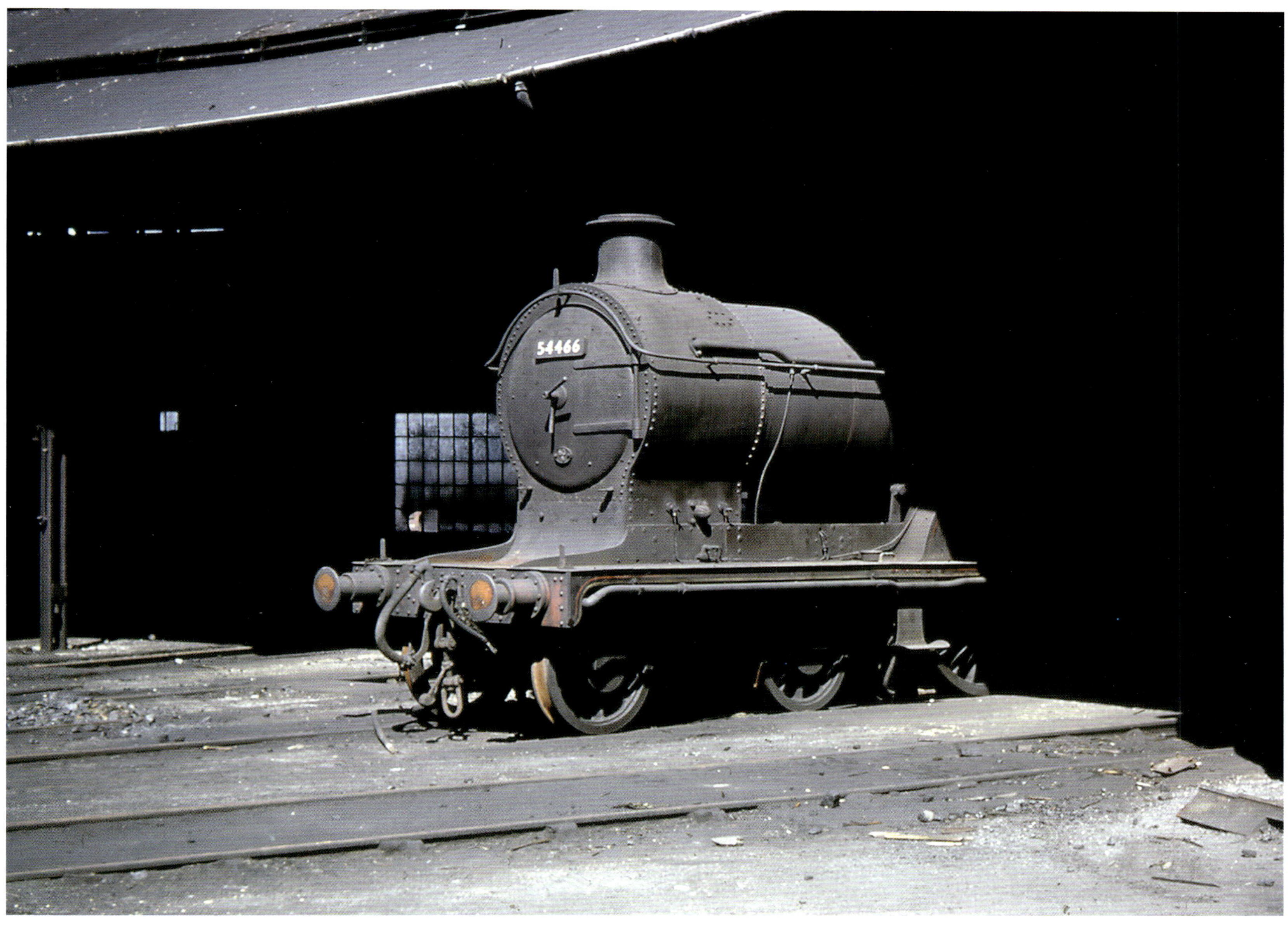

Sunday 3 June 1962. This solitary resident of Inverness roundhouse is ex-CR Class 113 (LMS Class 3P) 4-4-0 No. 54466. Withdrawn from service three months earlier, she had given forty-six years of service, having been constructed at St Rollox Works during 1916 and numbered 124 by the Caledonian Railway, later becoming No. 14466 with the LMS. (Hugh D. Ramsay)

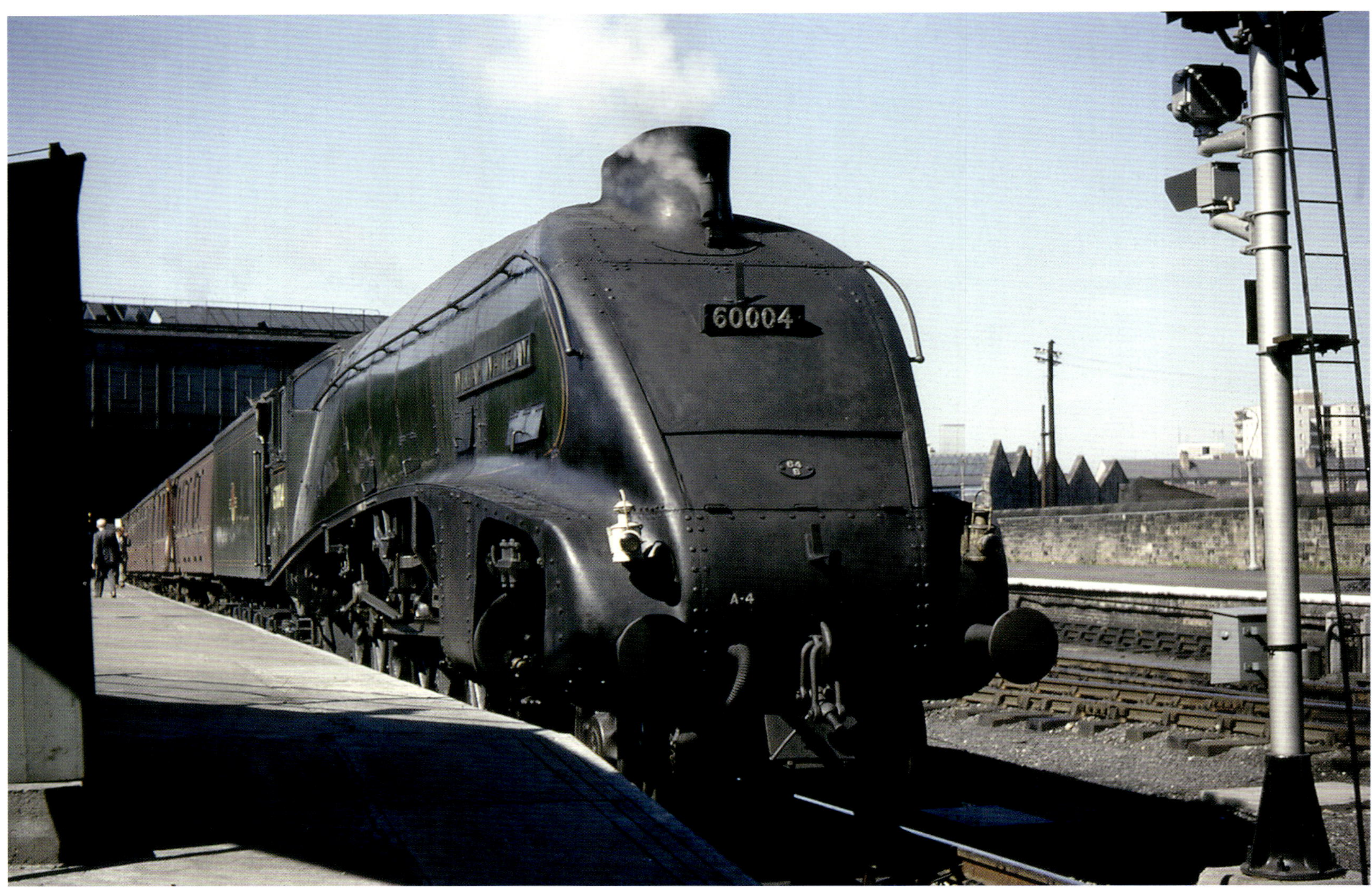

Tuesday 5 June 1962. Having arrived at Perth station at the head of an 'up' express from Aberdeen to Glasgow Buchanan Street, ex-LNER Class A4 4-6-2 No. 60004 *William Whitelaw* is waiting for the 'right of way'. Constructed at Doncaster Works during 1937 and originally named *Great Snipe*, she would be renamed after the chairman of the LNER in 1941. She is seen bearing a 64B Haymarket shed code but had just been re-allocated to Ferryhill in Aberdeen and would be withdrawn whilst there in 1966. (Hugh D. Ramsay)

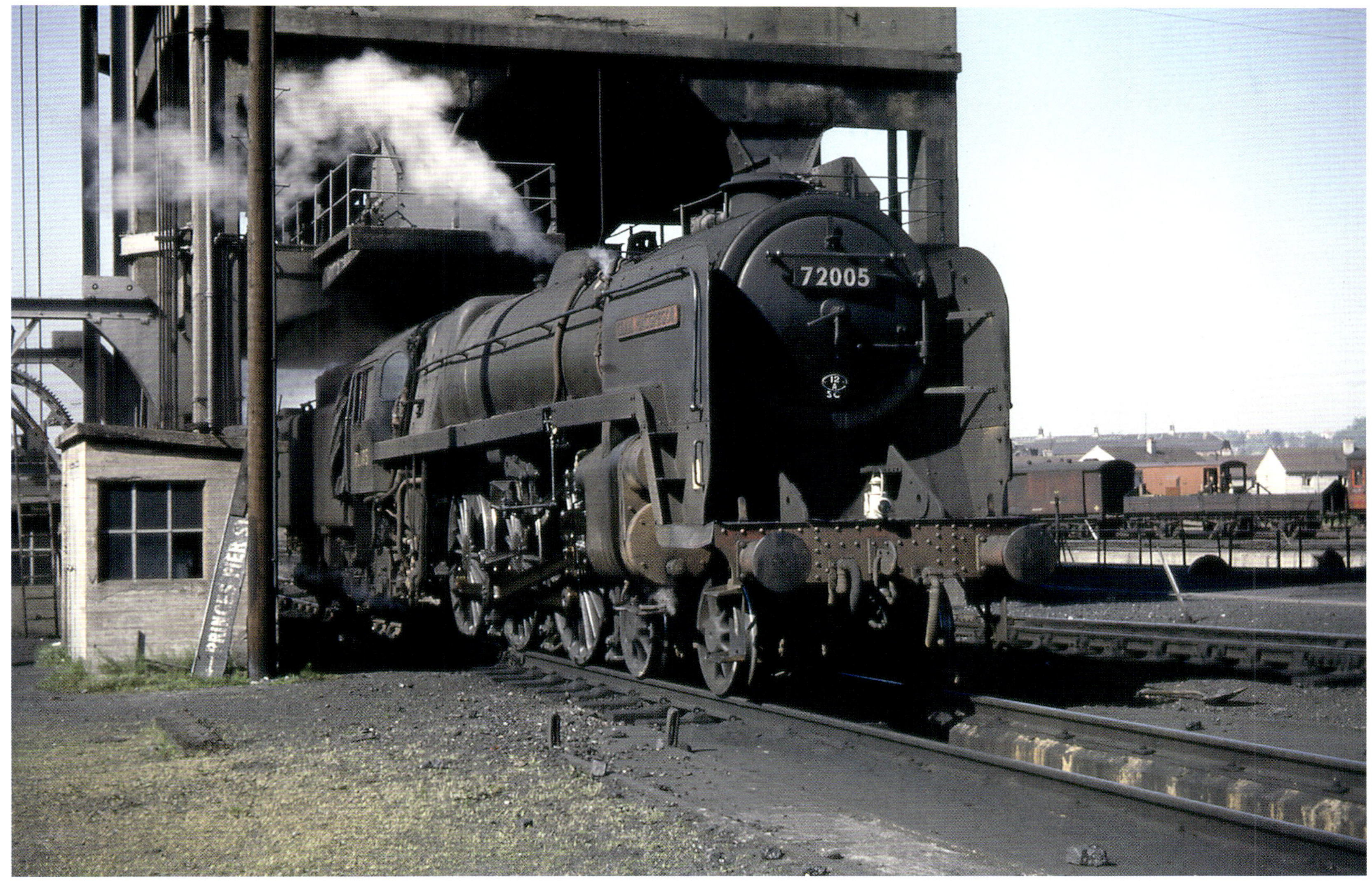

Tuesday 5 June 1962. Standing under the coaling tower at Perth shed is BR Standard Class 6P5F Clan 4-6-2 No. 72005 *Clan Macgregor*, bearing a 12A Carlisle Kingmoor shed code. One of only ten members of the class constructed, she entered service from Crewe Works during 1952, being allocated new to Kingmoor. She would end her days there, being withdrawn in 1965. Originally envisaged for use on the ex-Highland Main Line to Inverness, construction of the Clan Class was truncated after only ten examples were completed. They entered service between December 1951 and March 1952, all coming out of Crewe Works. Allocated equally between Glasgow Polmadie and Carlisle Kingmoor sheds, they saw much work on traffic between Glasgow, Manchester, Liverpool and on the 'Port Road' to Stranraer. The first five examples were withdrawn in 1962, after only eleven years of service, the remainder going for scrap during 1965 and 1966. (Hugh D. Ramsay)

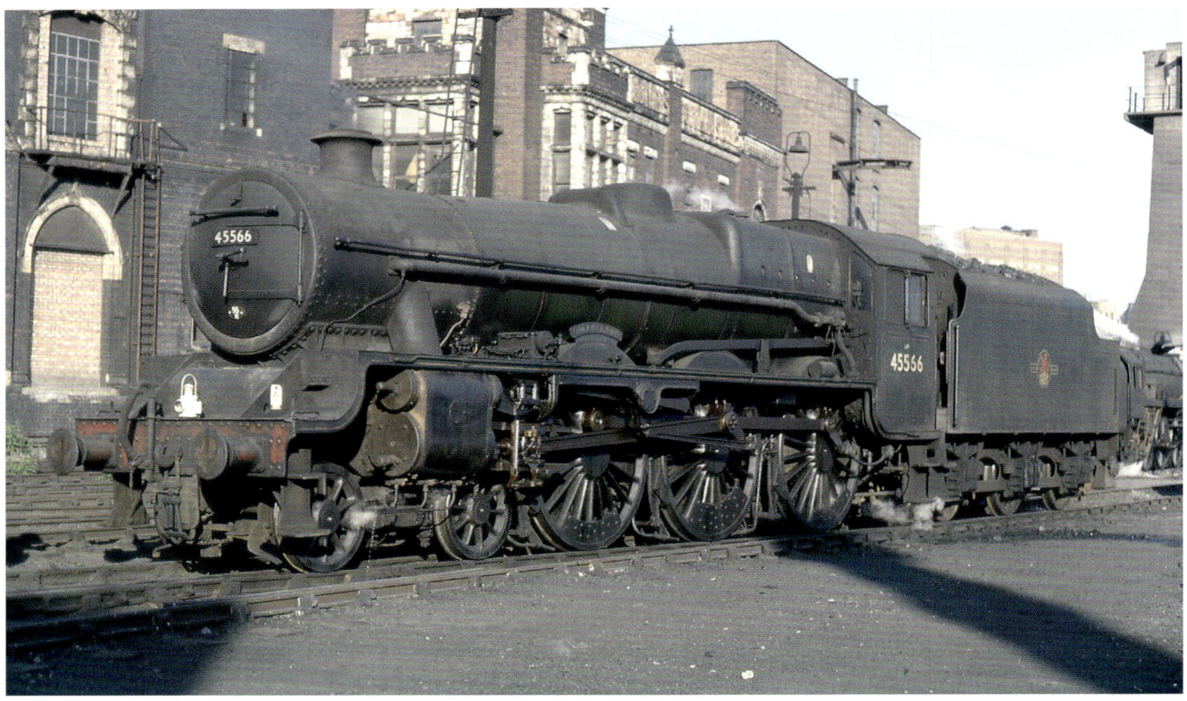

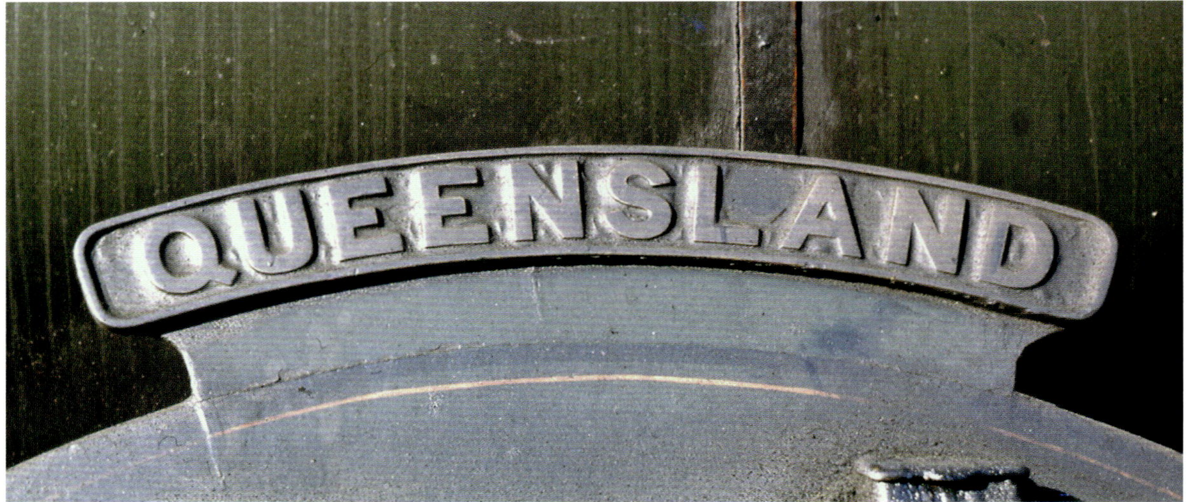

Thursday 14 June 1962. *Top:* At Kentish Town shed ex-LMS Class 6P5F Jubilee 4-6-0 No. 45566 *Queensland* is standing in the yard. Allocated to Leeds Holbeck shed and carrying the correct 55A shed code, she had been constructed by the NBL during 1934 and would be withdrawn from service five months after this photograph was taken, in November 1962. *Bottom:* The nameplate of *Queensland* on the driver's side of the locomotive. (Both Andrew G. Forsyth)

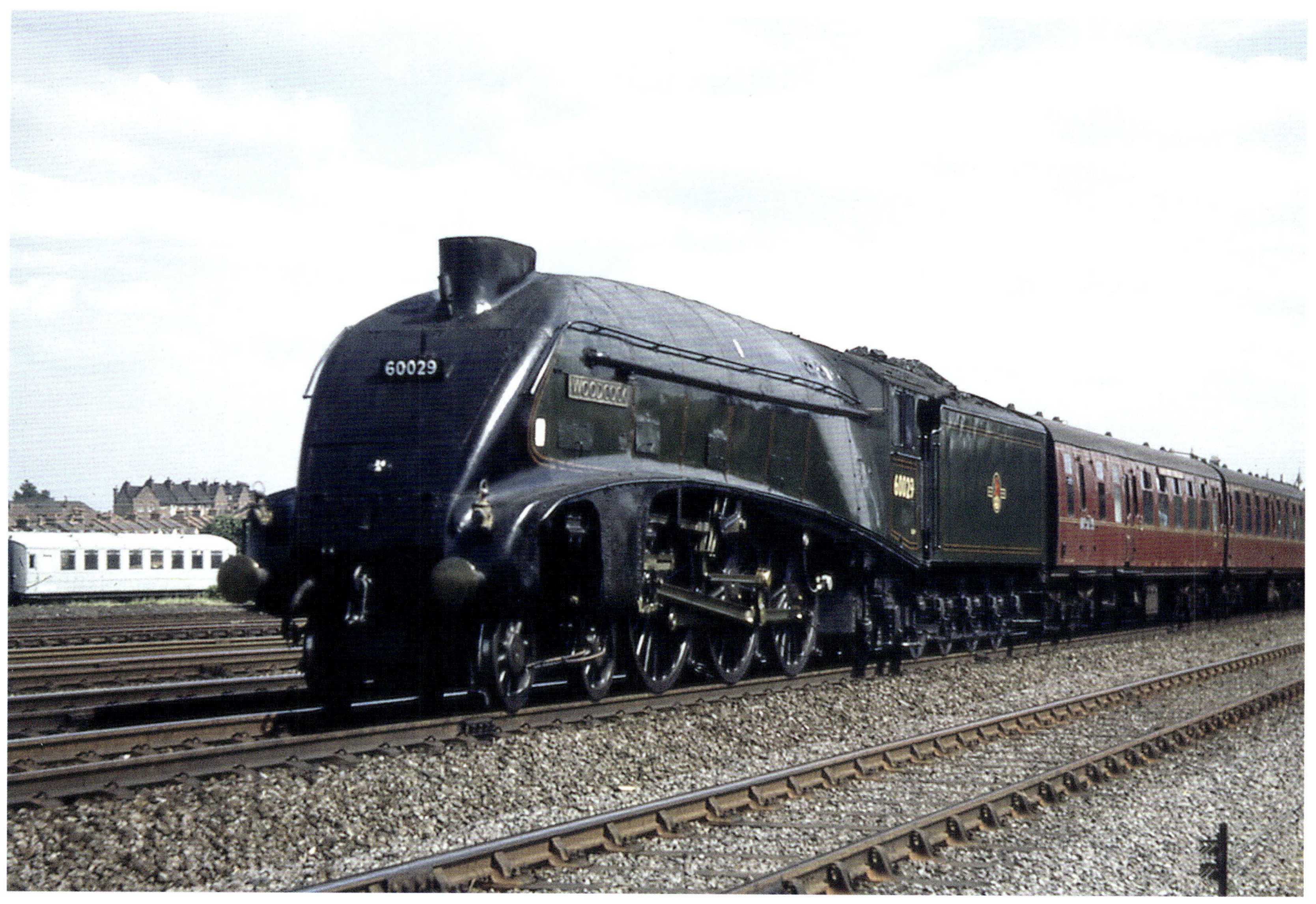

Monday 18 June 1962. In sparking condition, ex-LNER Class A4 4-6-2 No. 60029 *Woodcock* is at the head of a 'down' express just north of New Southgate station. Constructed at Doncaster Works during 1937, she was allocated new to Gateshead shed but spent much of her working life based at Kings Cross shed, being withdrawn in 1963. (Hugh D. Ramsay)

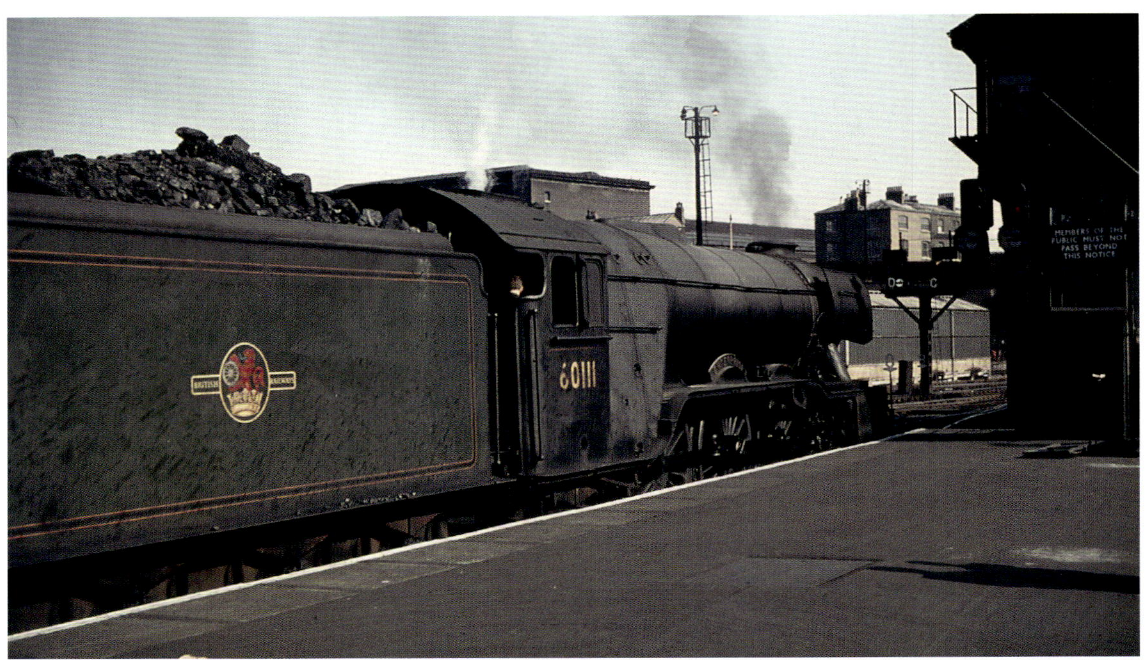

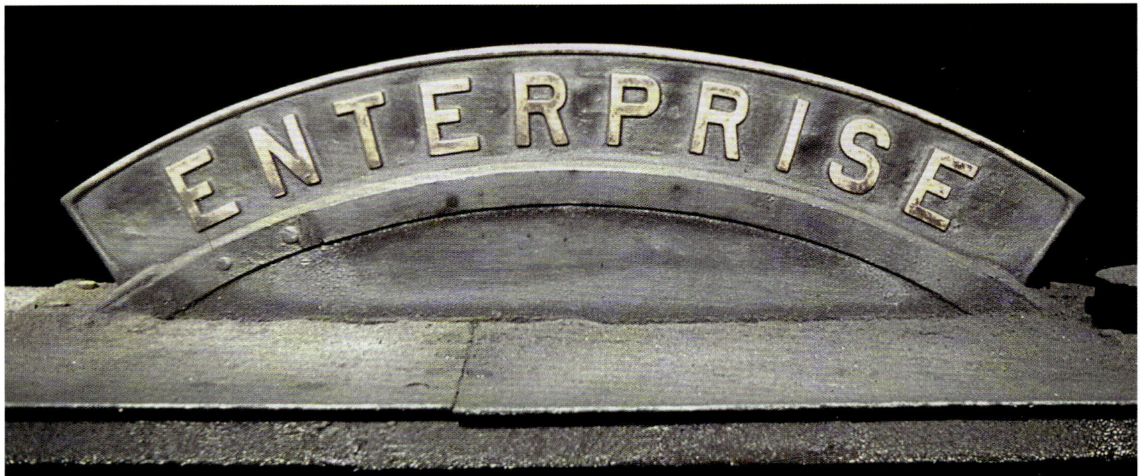

Saturday 7 July 1962. Standing at the head of an 'up' express at platform six in Kings Cross station is ex-LNER Class A3 4-6-2 No. 60111 *Enterprise*. Constructed at Doncaster Works during 1923 as a Class A1 locomotive, she would be rebuilt as a Class A3 locomotive in 1927 and acquire a double chimney in 1959. Being fitted with trough-style smoke deflectors early in 1962, she would be withdrawn five months after this photograph was taken, during December 1962. *Top:* No. 60111 awaiting departure at the head of its train. *Bottom:* The fireman's side nameplate *Enterprise*, the locomotive so named after the racehorse that won the 1887 2,000 Guineas race. (Both Hugh D. Ramsay)

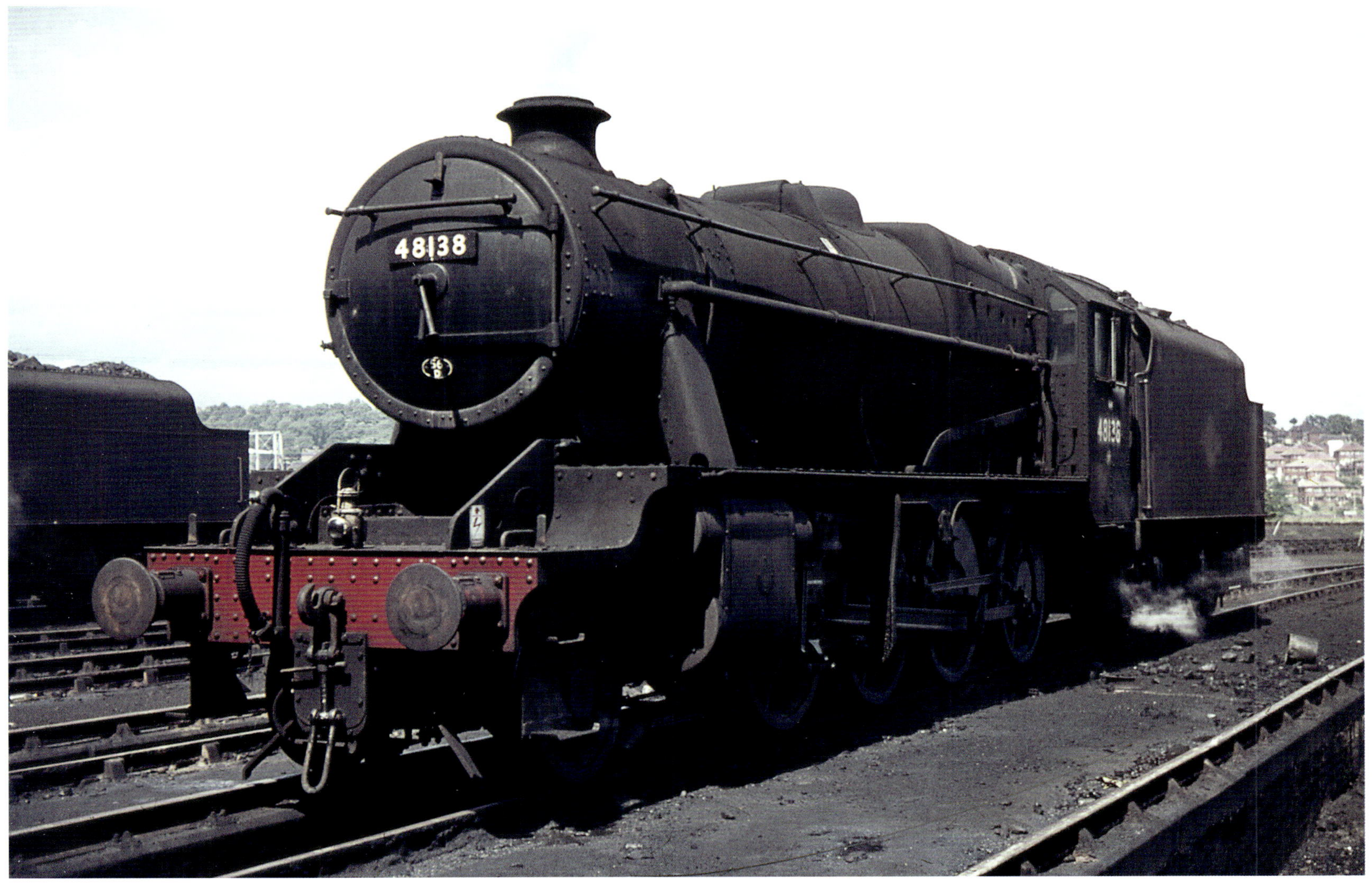

Saturday 21 July 1962. Bearing a 56D Mirfield shed code, ex-LMS Class 8F 2-8-0 No. 48138 is seen standing at the former L&NWR Farnley Junction shed. Constructed at Crewe Works in 1941, she would be allocated to Mirfield during 1960 and withdrawn whilst based there in 1965. The William Stanier Class 8F 2-8-0 was one of the most successful designs of heavy goods locomotives produced by any of the post-grouping companies. A total of 852 examples were constructed between 1935 and 1946 by no less than eight railway works – Ashford, Brighton, Crewe, Darlington, Doncaster, Eastleigh, Horwich and Swindon – which produced a total of 525 members of the class. A further 327 were constructed by independent contractors Beyer Peacock & Co., the North British Locomotive Co. and the Vulcan Foundry. Many saw service during and after the Second World War with railways in Italy, Iran, Iraq, Turkey, Palestine and Israel. A total of fourteen members of the class have survived, with eight in the UK and six in Turkey and Iraq. (Hugh D. Ramsay)

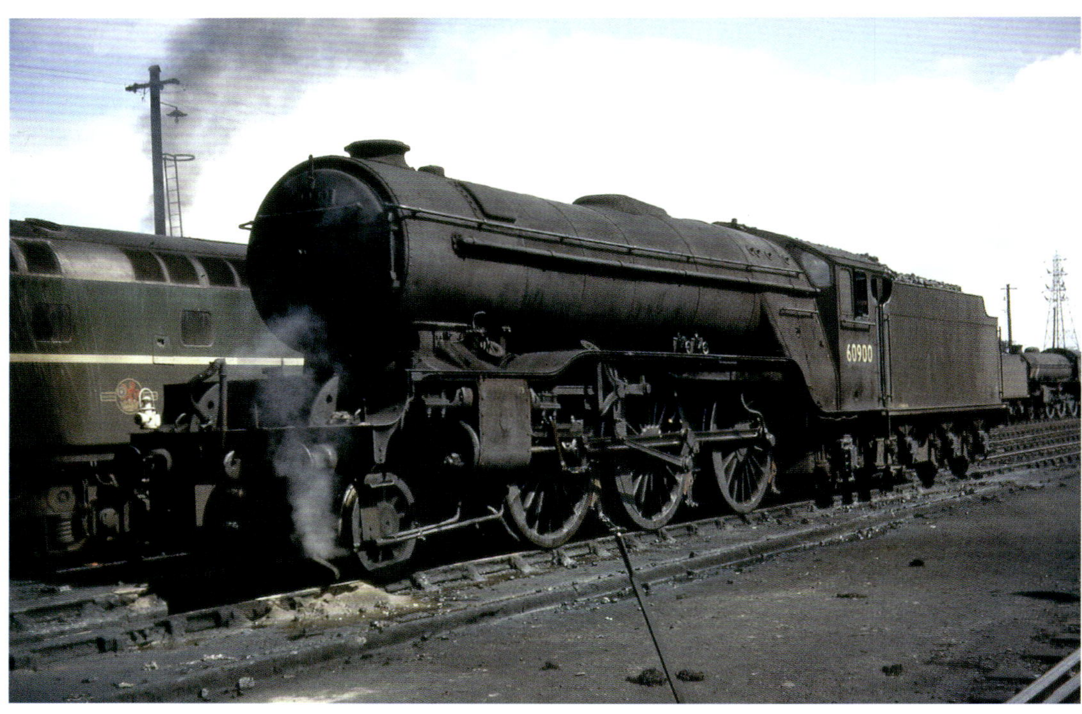

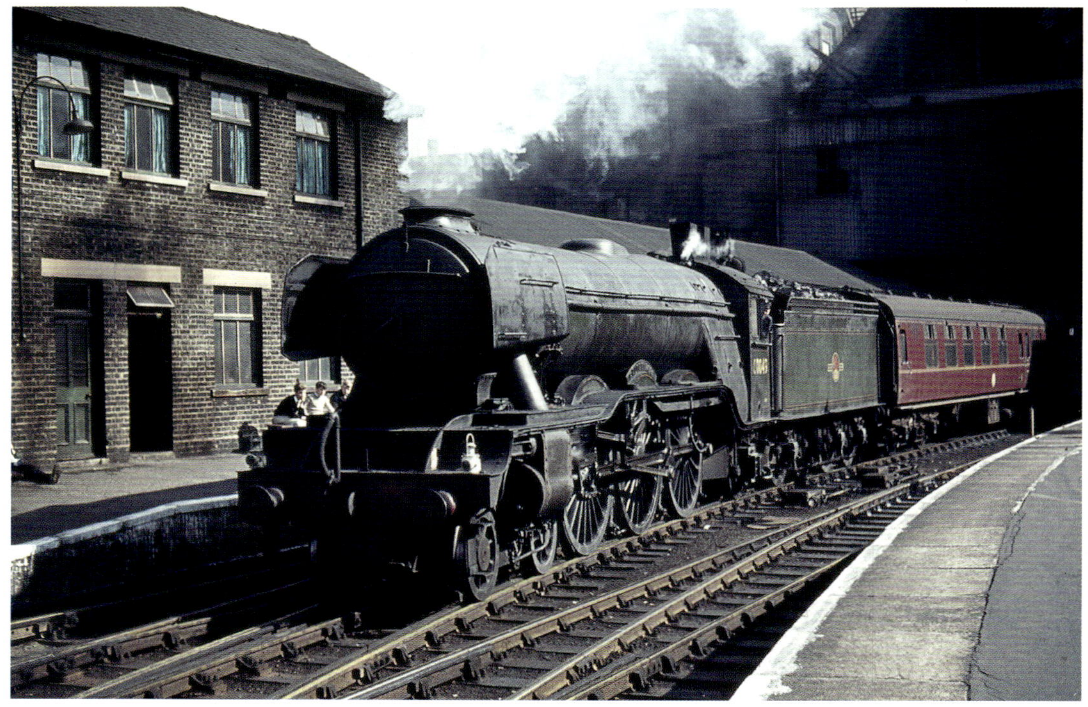

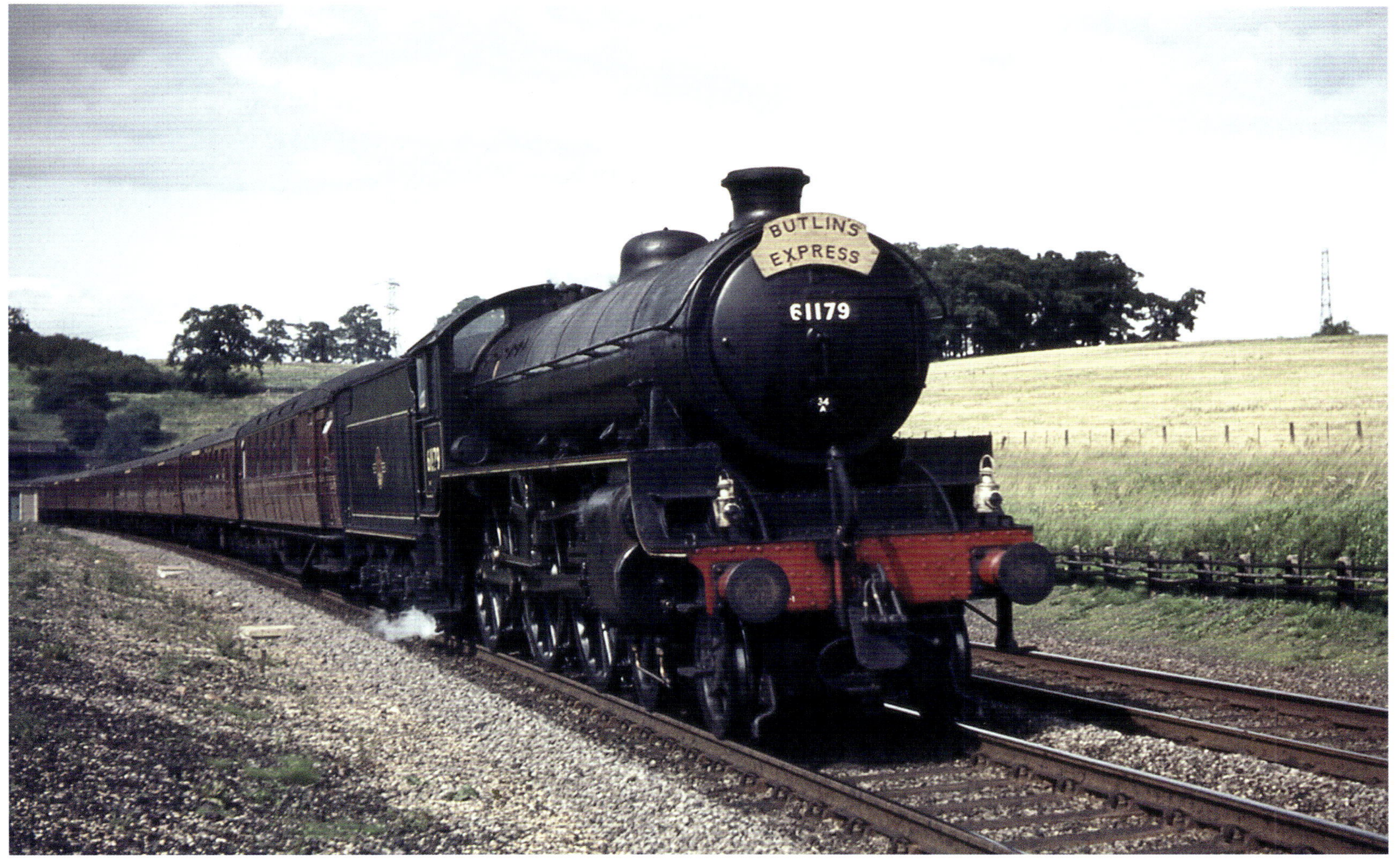

Opposite top: Monday 6 August 1962. Seen at Carlisle Canal shed is ex-LNER Class V2 2-6-2 No. 60900. Constructed at Darlington Works during 1940, she would be allocated new to Kings Cross and, during 1953, be re-allocated to St Margarets in Edinburgh. She would be withdrawn from service during 1963. (Hugh D. Ramsay)

Opposite bottom: Friday 17 August 1962. Seen departing from platform six at Kings Cross station at the head of an express is ex-LNER Class A3 4-6-2 No. 60049 *Galtee More*. Entering service from Doncaster Works in 1928 as a Class A1 locomotive, she would be rebuilt during 1945 as a Class A3. Acquiring a double chimney in 1959 and trough-style smoke deflectors in 1960, she would be withdrawn from service four months after this photograph was taken, in December 1962. She was named after the racehorse that won the Derby, St Leger and 2,000 Guineas in 1897. (Hugh D. Ramsay)

Above: Saturday 25 August 1962. At Ganwick curve, between Potters Bar and Hadley Wood, ex-LNER Class B1 4-6-0 No. 61179 has just emerged from the tunnel at the head of an 'up' 'Butlin's Express' from Skegness, due to arrive at Kings Cross at 3.23 p.m. Entering service from the Vulcan Foundry in 1947, she would be allocated to Kings Cross during 1958, and here is bearing the correct 34A shed code. She would end her days based at Immingham, being withdrawn in 1965. (Hugh D. Ramsay)

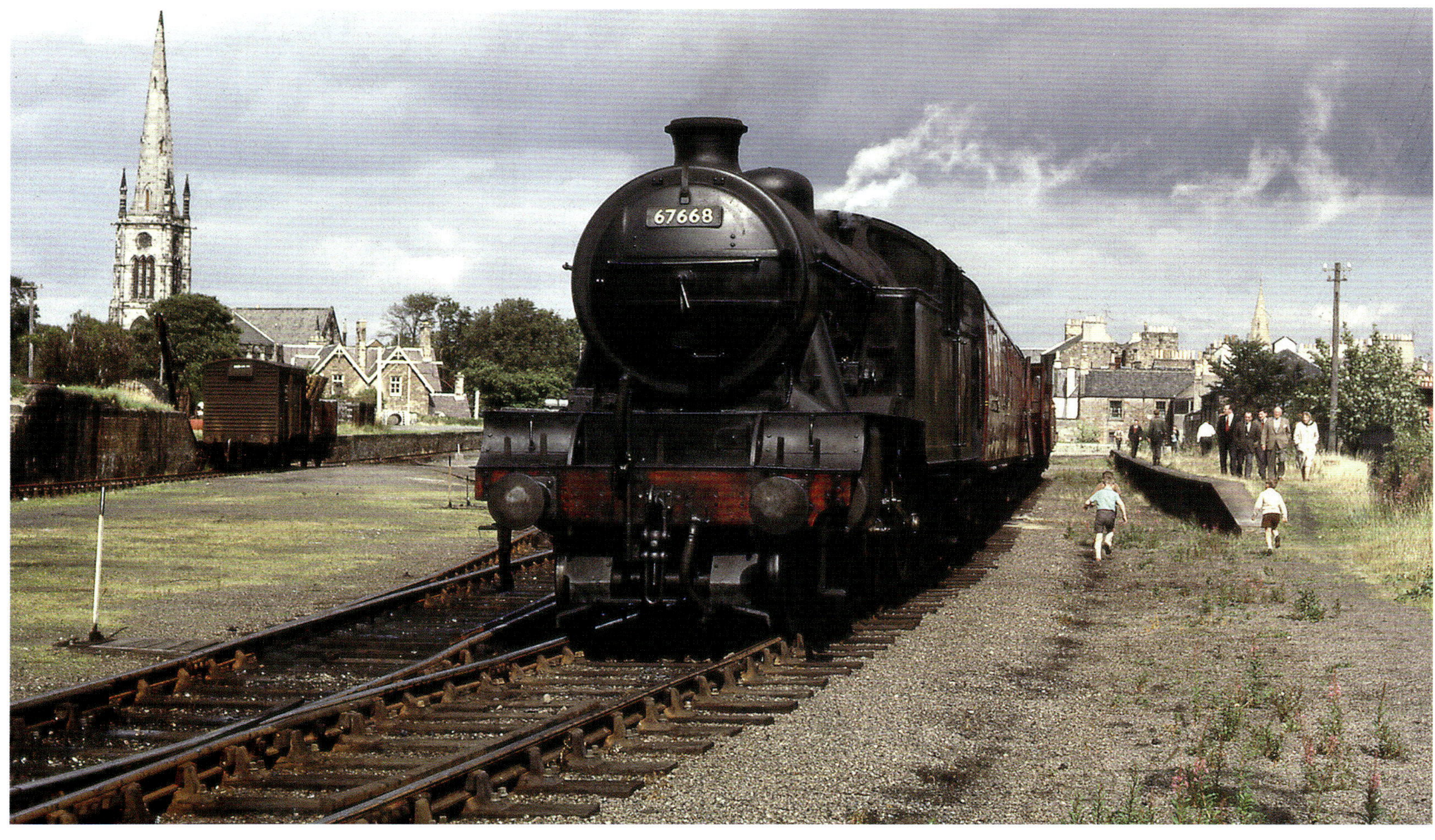

Above: Saturday 25 August 1962. During the afternoon of this day, the SLS Scottish area ran the 'Edinburgh and Dalkeith Rail Tour', commencing at Waverley station and visiting the former Edinburgh and Dalkeith Railway terminus at Dalkeith. The train then proceeded to St Leonards goods depot in the city, Leith South and Musselburgh before finally returning to Waverley station in the evening. Motive power included ex-LNER Class V3 2-6-2 tank No. 67668, which is seen here at Dalkeith. Constructed at Doncaster Works during 1938 as a Class V1 locomotive, she would be rebuilt in 1954 as a Class V3 and would be withdrawn from service four months after this photograph was taken, during December 1962. The impressive tower and spire in the background belong to the West Church in Dalkeith. (Andrew G. Forsyth)

Opposite top: Sunday 26 August 1962. At St Margarets shed in Edinburgh, ex-LNER Class A3 4-6-2 No. 60089 *Felstead* is seen in grimy, work-stained condition and is bearing the correct 64A shed code. Entering service from Doncaster Works during 1928, she would be fitted with a double chimney in 1959 and trough-style smoke deflectors in 1961, and then withdrawn from service during 1963. She carries the name of the racehorse that won the 1928 Derby race. (Andrew G. Forsyth)

Opposite bottom: Tuesday 28 August 1962. Waiting for passengers to board at Perth station, BR Standard Class 5 4-6-0 No. 73120 is at the head of a Glasgow-bound stopper, which contains two meat container wagons behind its tender. A crew member is standing on the platform waiting for 'the right away' from the guard. Constructed at Doncaster Works during 1956, she was allocated new to Perth shed but would end her days based at Corkerhill in Glasgow, being withdrawn during 1966. (Andrew G. Forsyth)

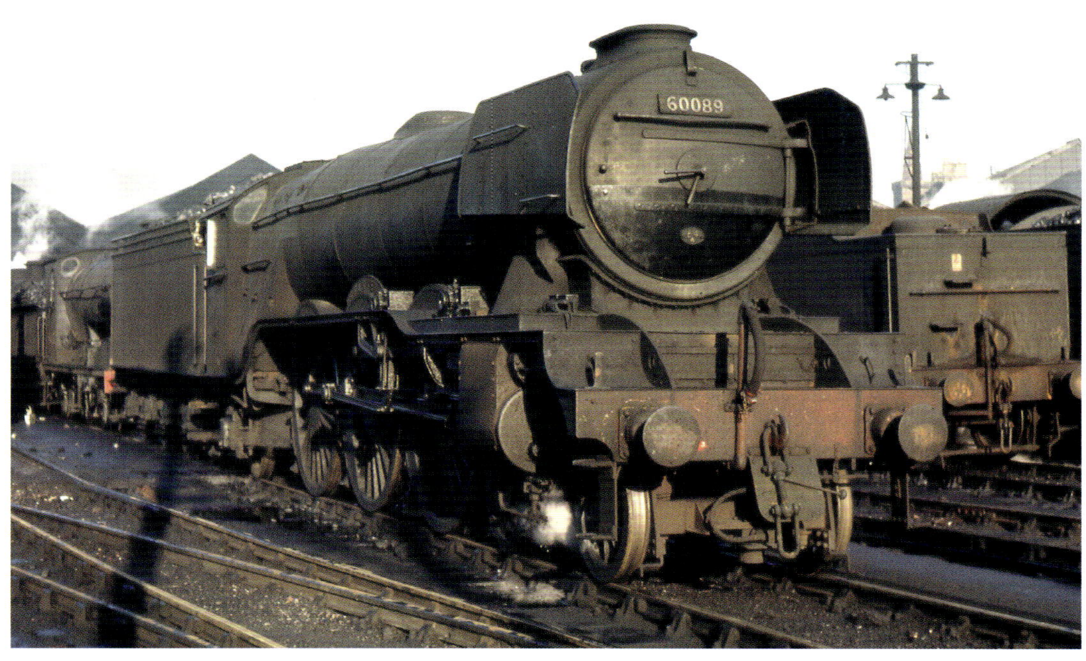

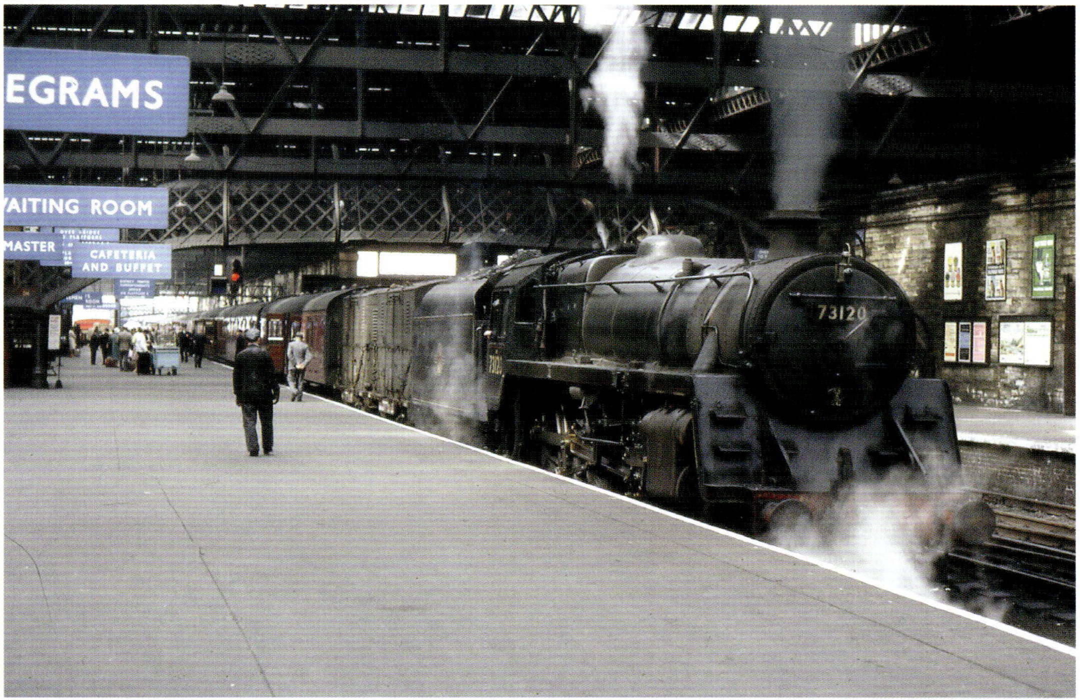
EGRAMS
AITING ROOM
MASTER
CAFETERIA
AND BUFFET
73120

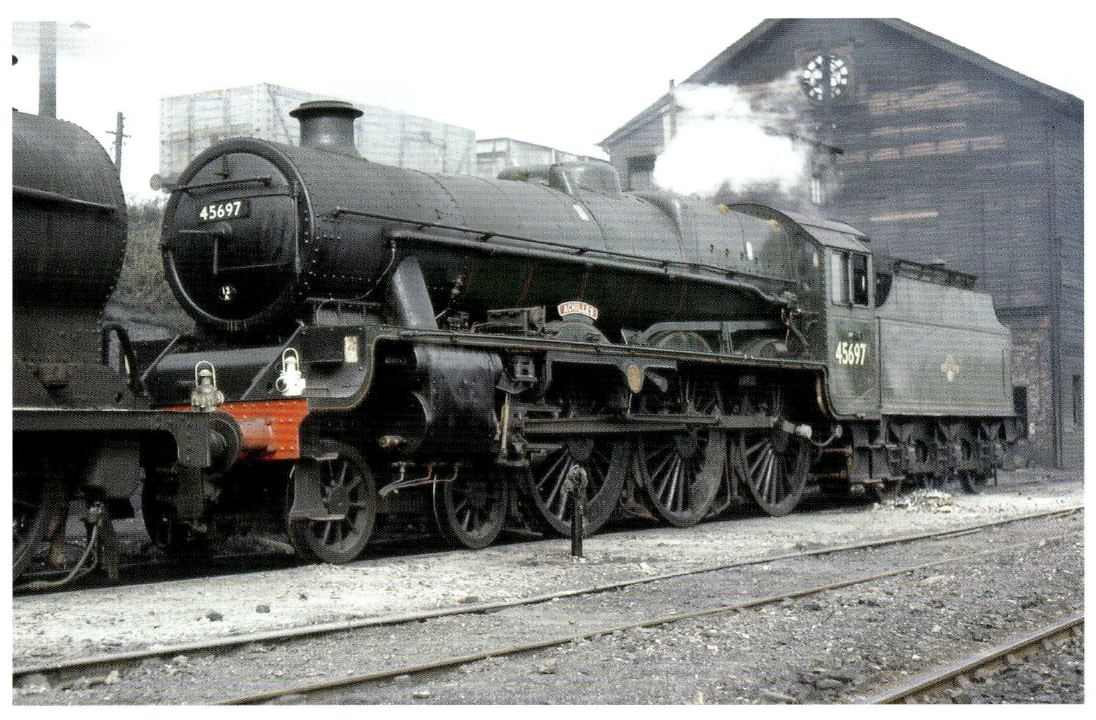

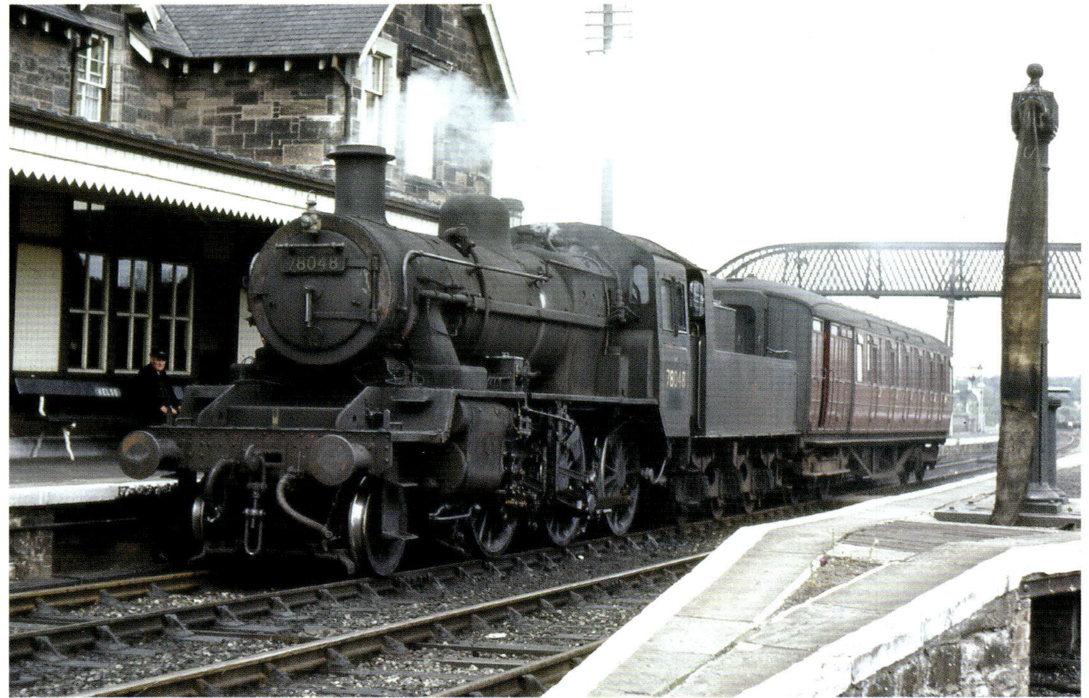

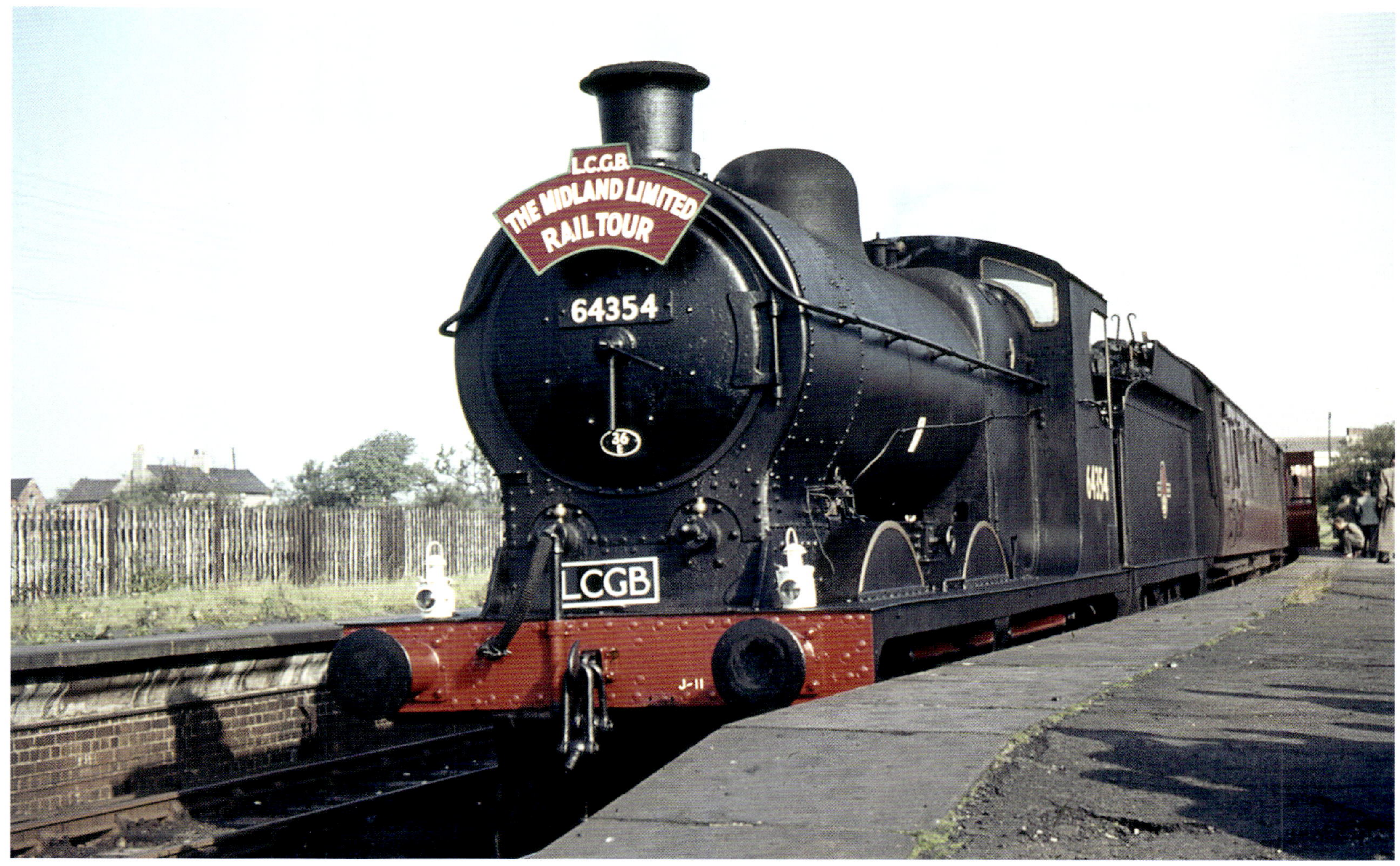

Opposite top: Wednesday 29 August 1962. This ex-LMS Class 6P5F Jubilee 4-6-0 No. 45697 *Achilles* is seen at what the photographer noted is Balornock shed in Glasgow. This shed was originally built by the Caledonian Railway during 1916 but became referred to as St Rollox by the LMS. The Jubilee was a product of Crewe Works during 1936 and is seen carrying a 12A Carlisle Kingmoor shed code. She would end her days based at Holbeck in Leeds and be withdrawn in 1967. (Andrew G. Forsyth)

Opposite bottom: Thursday 30 August 1962. At Kelso station on the Tweed Valley Line, the driver of BR Standard Class 2 2-6-0 No. 78048 is seen sitting on the platform bench while waiting for departure to St Boswells of his single-coach train. Constructed at Darlington Works in 1955 and allocated

new to St Margarets in Edinburgh, she would by 1960 be allocated to Hawick and withdrawn from service during 1964. (Andrew G. Forsyth)

Above: Sunday 14 October 1962. This day saw the running of the LCGB 'Midland Limited Rail Tour', commencing at Marylebone station and running via Aylesbury, Rugby Central, Nottingham Victoria, visiting Burton-on-Trent and returning to London via Derby, Leicester, Bedford and ending at St Pancras station. Motive power included ex-GCR Class 9J (LNER Class J11) 0-6-0 No. 64354, which handled the train between Nottingham and Burton-on-Trent, and is seen here at Egginton Junction. Entering service from Gorton Works in 1903, she would be withdrawn during the same month as this photograph was taken. (Hugh D. Ramsay)

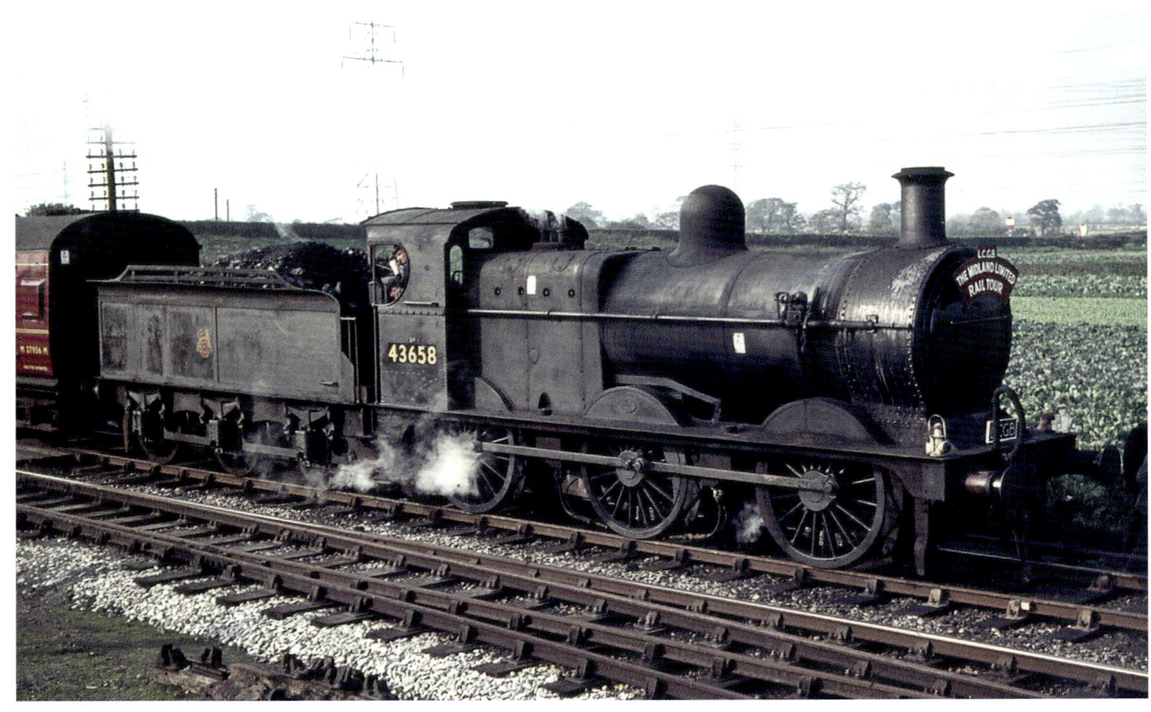

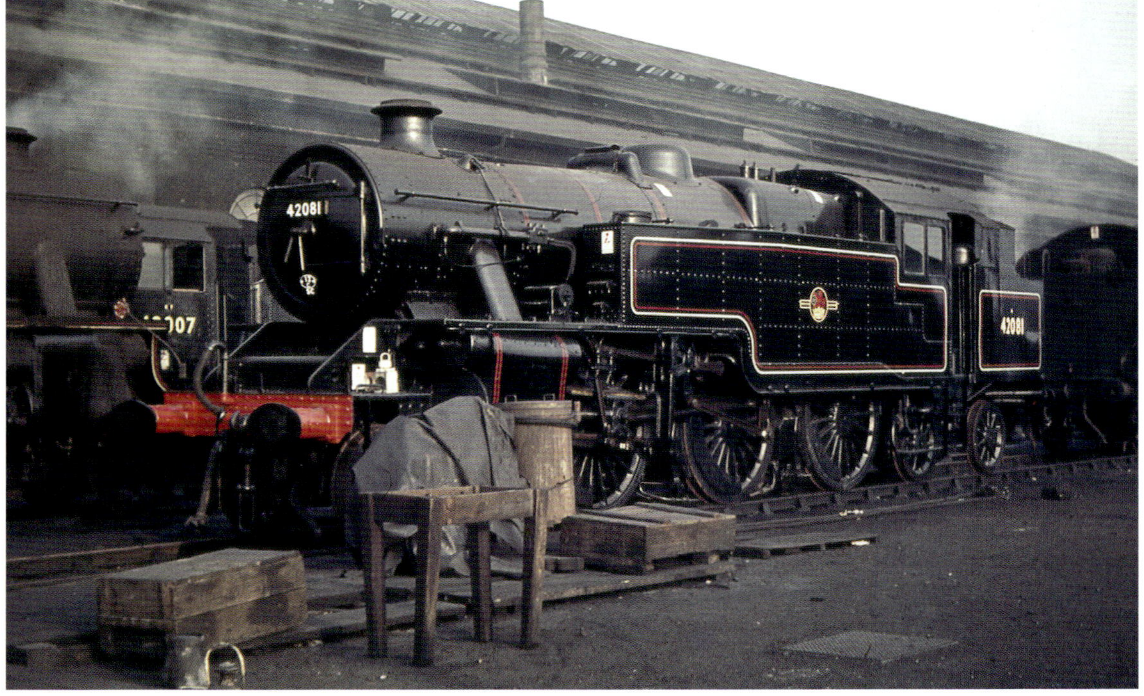

Opposite top: Sunday 14 October 1962. Hauling the section between Burton-on-Trent and Derby Midland was the responsibility of ex-MR Class 1873 (LMS Class 3F) 0-6-0 No. 43658, which is seen here at the Castle Donnington stop. Constructed at the Vulcan Foundry in 1900, she would be withdrawn from service during 1963. (Hugh D. Ramsay)

Opposite bottom: Sunday 14 October 1962. Seen in ex-works condition after an overhaul at Derby Works and waiting to be run-in is Class 4P 2-6-4 tank No. 42081. Constructed at Brighton Works during 1951, she would be allocated to Stewarts Lane shed in London but would become something of a wanderer, spending time based at Neasden, Bangor, Carlisle Canal and Blackpool, finally being withdrawn in 1967 while based at Trafford Park. (Hugh D. Ramsay)

Right: Sunday 10 March 1963. This close-up of the nameplate of ex-LNER Class A3 4-6-2 No. 60071 *Tranquil* was taken at Kings Cross shed. Constructed by the NBL in 1924 as a Class A1 locomotive, she would be rebuilt during 1944 as a Class A3. Allocated new to Gateshead shed, she would spend time based at other North Eastern Region sheds and withdrawn in 1964. She was named after the racehorse that had won the 1923 St Leger and 1,000 Guineas races. (Hugh D. Ramsay)

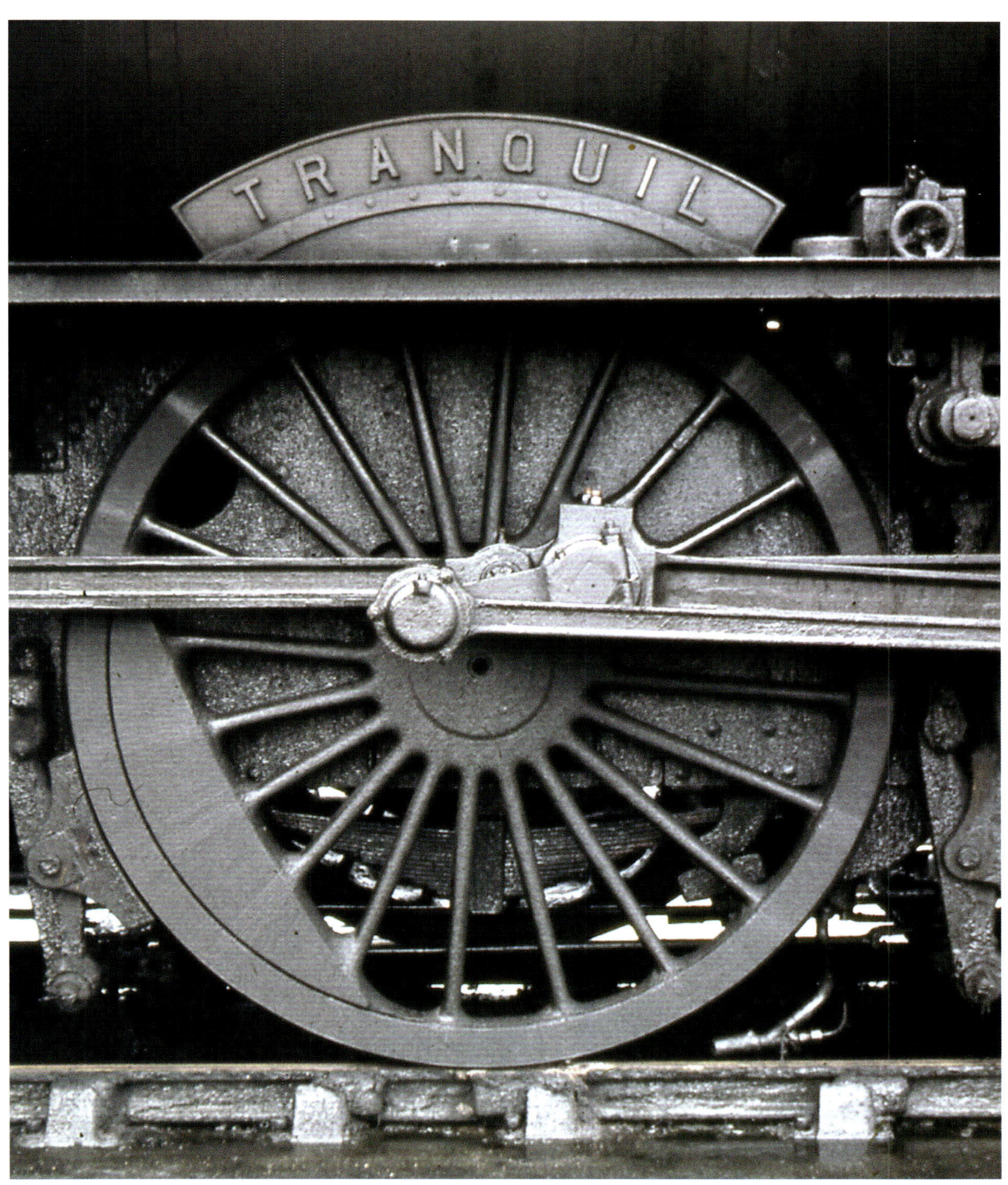

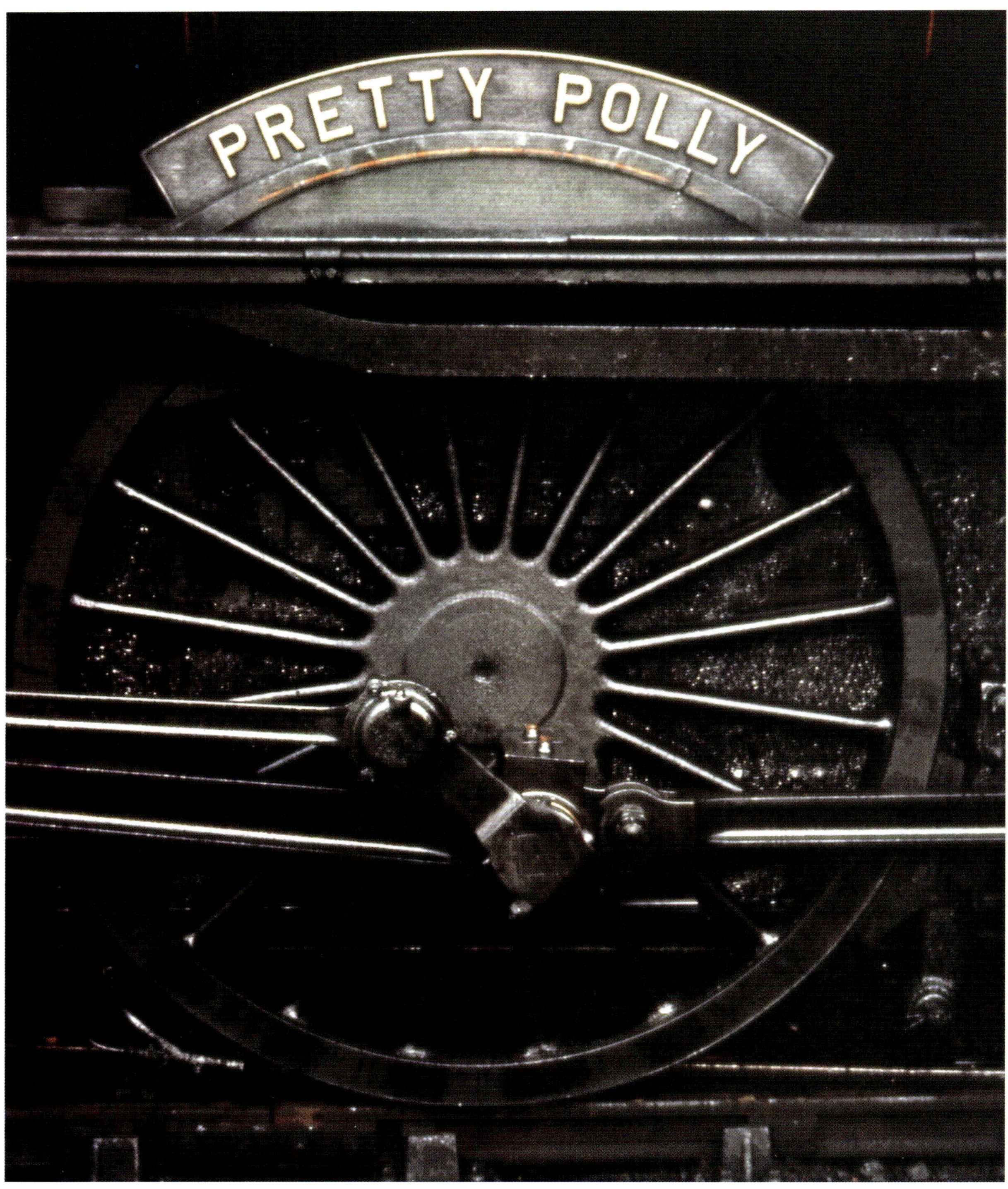

Wednesday 13 March 1963. This close-up of the nameplate of ex-LNER Class A3 4-6-2 No. 60061 *Pretty Polly* was taken at Kings Cross station. Constructed at Doncaster Works in 1925 as a Class A1 locomotive, she would be rebuilt in 1944 as a Class A3. Allocated new to Gorton shed, she would also spend time based at Kings Cross before being withdrawn during 1963. (Hugh D. Ramsay)

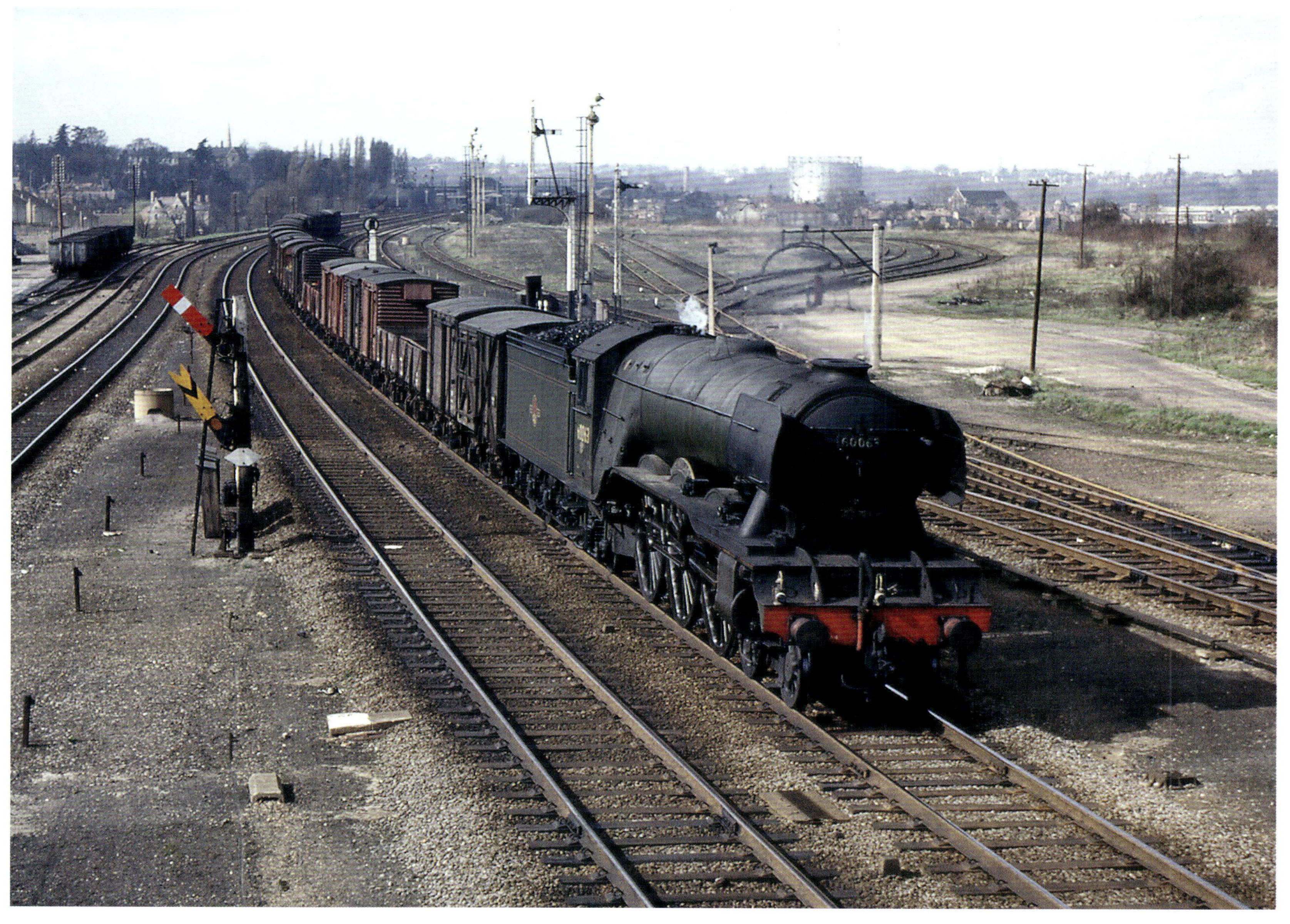

Saturday 4 May 1963. Seen approaching Oakleigh Park station at the head of an 'up' goods train is ex-LNER Class A3 4-6-2 No. 60063 *Isinglass*. Constructed at Doncaster Works during 1925 as a Class A1 locomotive, she would be rebuilt as a Class A3 in 1946. Acquiring a double chimney in 1959 and trough-style smoke deflectors in 1961, she would be withdrawn from service in 1964. She bore the name of the racehorse that won the 1893 Derby, St Leger and 2,000 Guineas races. (Andrew G. Forsyth)

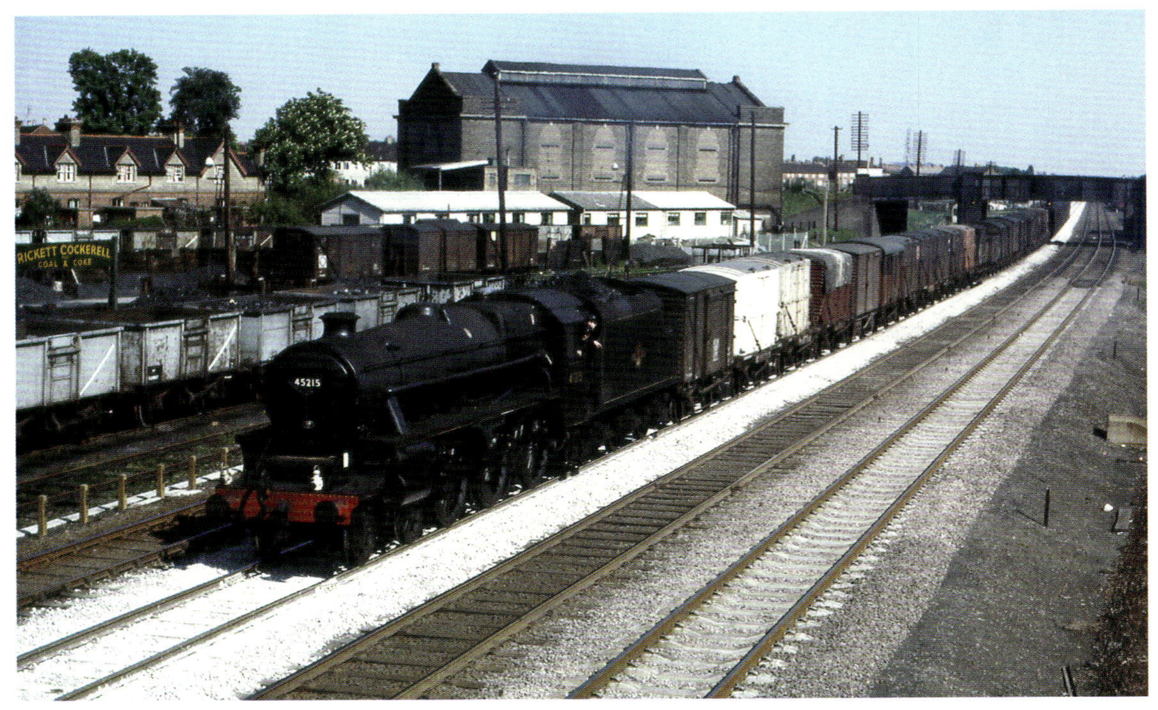

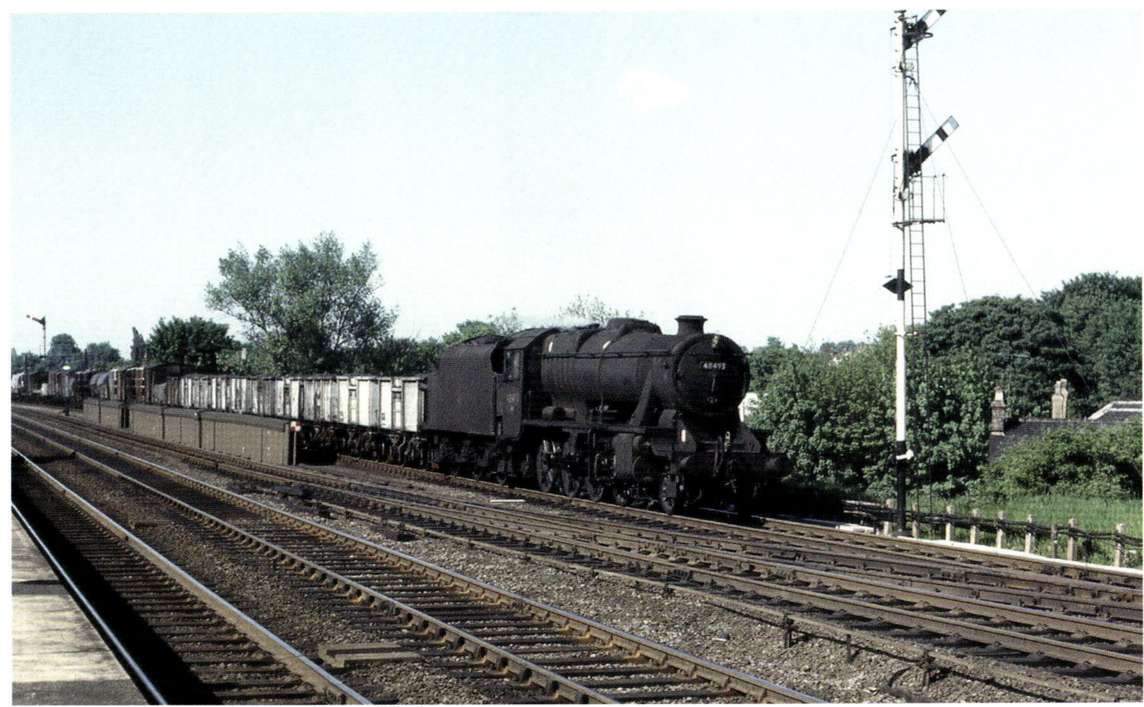

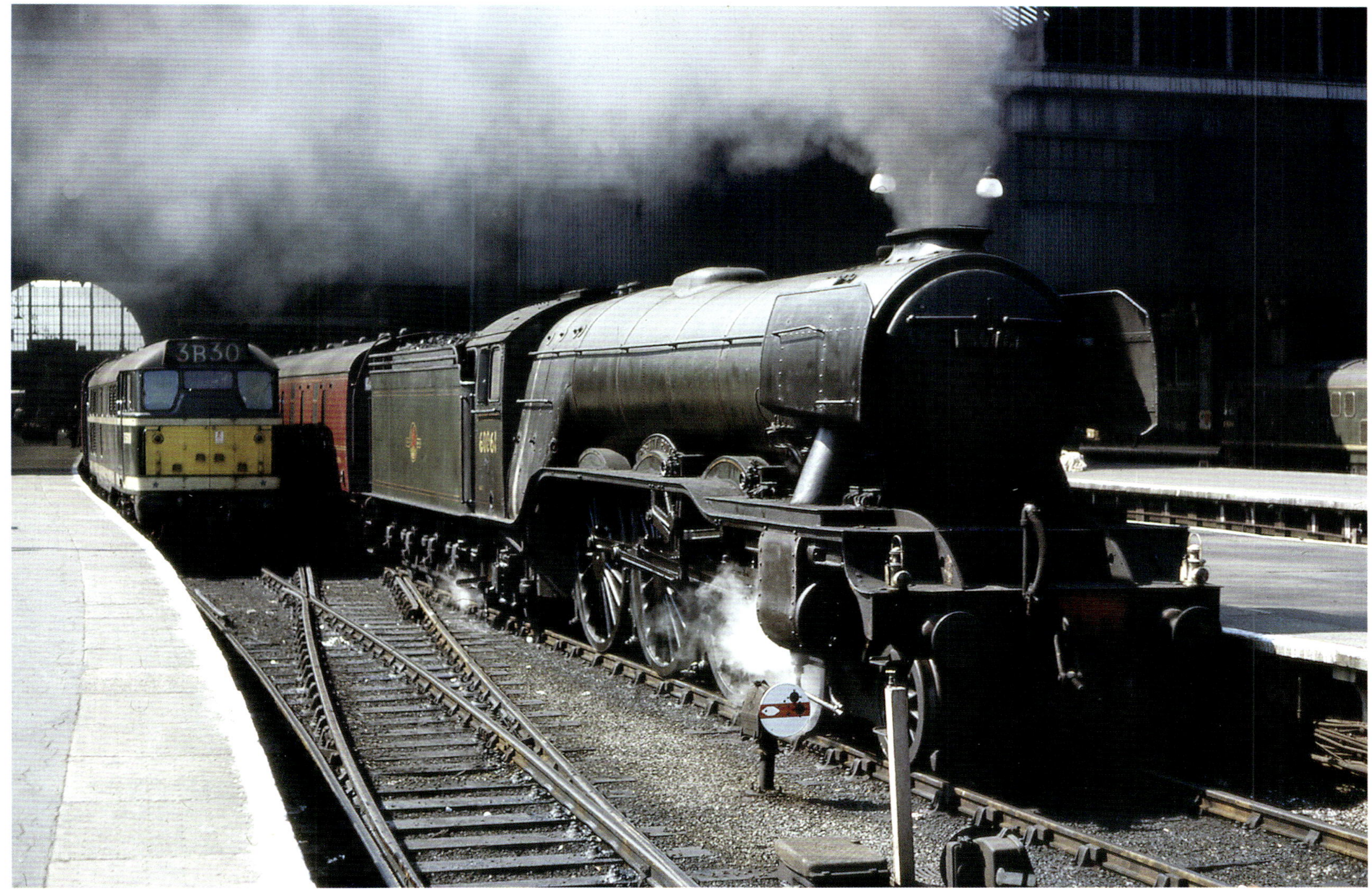

Opposite top: Friday 1 June 1963. Heading north at the head of a long goods train and about to pass Headstone Lane station is ex-LMS Class 5 4-6-0 No. 45215, looking in good, clean condition. Entering service from Armstrong Whitworth & Co. during 1935, she would give thirty-two years of service before being withdrawn in 1967. (Andrew G. Forsyth)

Opposite bottom: Saturday 2 June 1963. At Bushey and Oxhey station, ex-LMS Class 8F 2-8-0 No. 48493 is seen heading a 'down' mixed-goods train. Constructed at Horwich Works in 1945, she would be withdrawn during 1968. (Andrew G. Forsyth)

Above: Saturday 8 June 1963. Departing from Kings Cross station at the head of an express is ex-LNER Class A3 4-6-2 No. 60061 *Pretty Polly*. A product of Doncaster Works as a Class A1 locomotive during 1925, she would be rebuilt as a Class A3 in 1944. Fitted with a double chimney in 1958 and trough-style smoke deflectors in 1962, she would be withdrawn three months after this photograph was taken, during September 1963. (Andrew G. Forsyth)

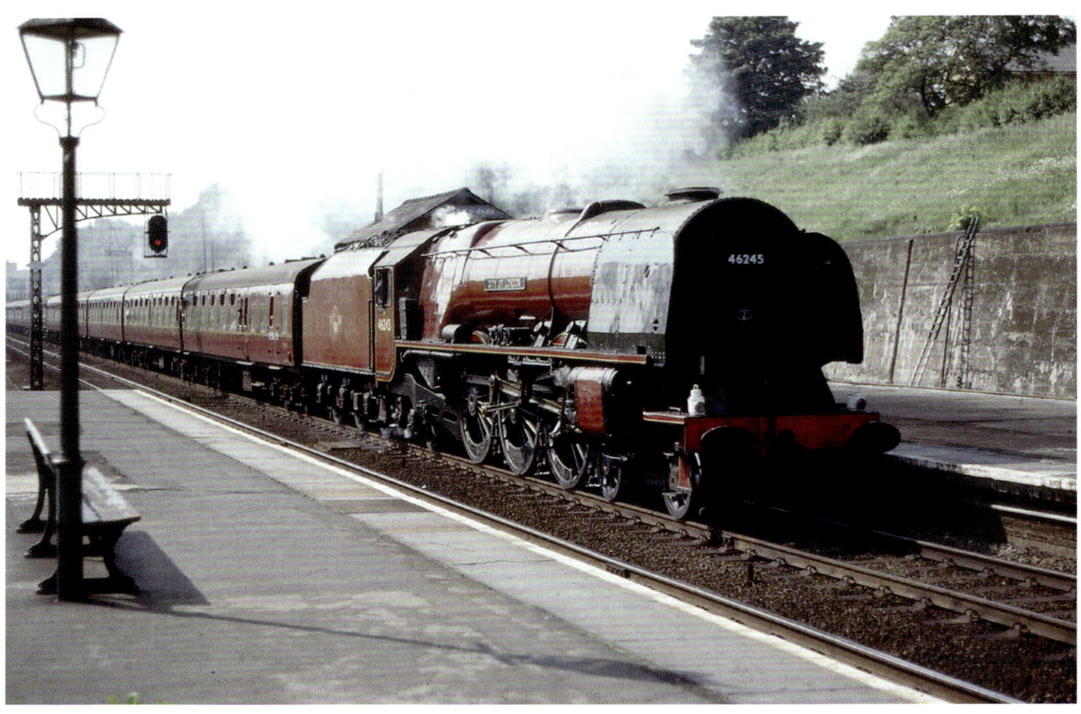

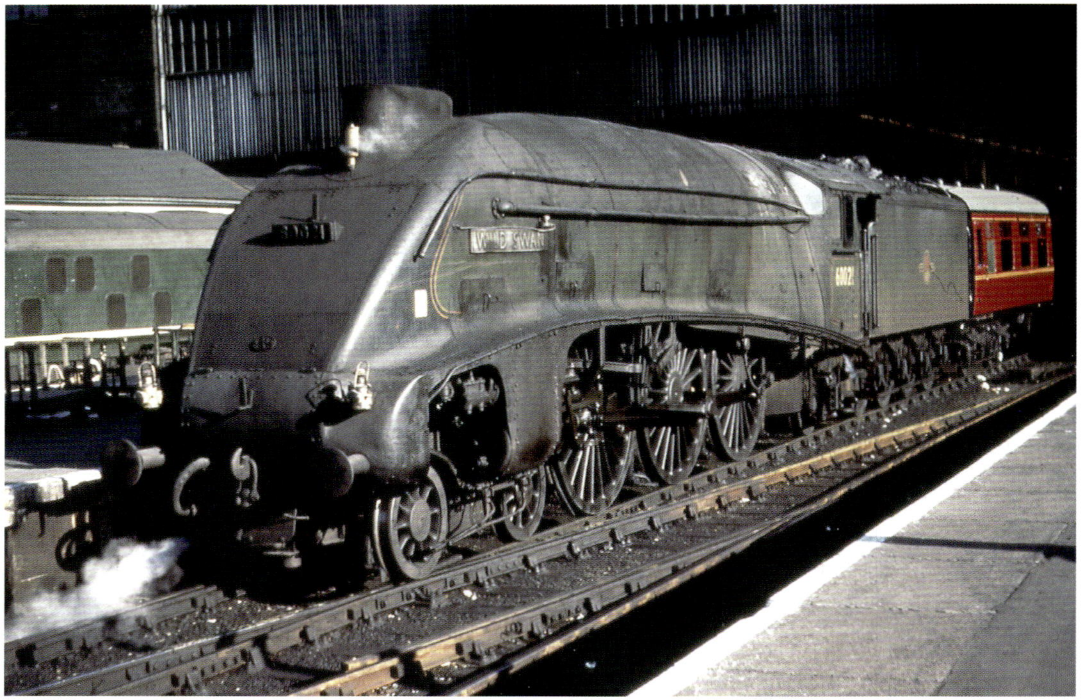

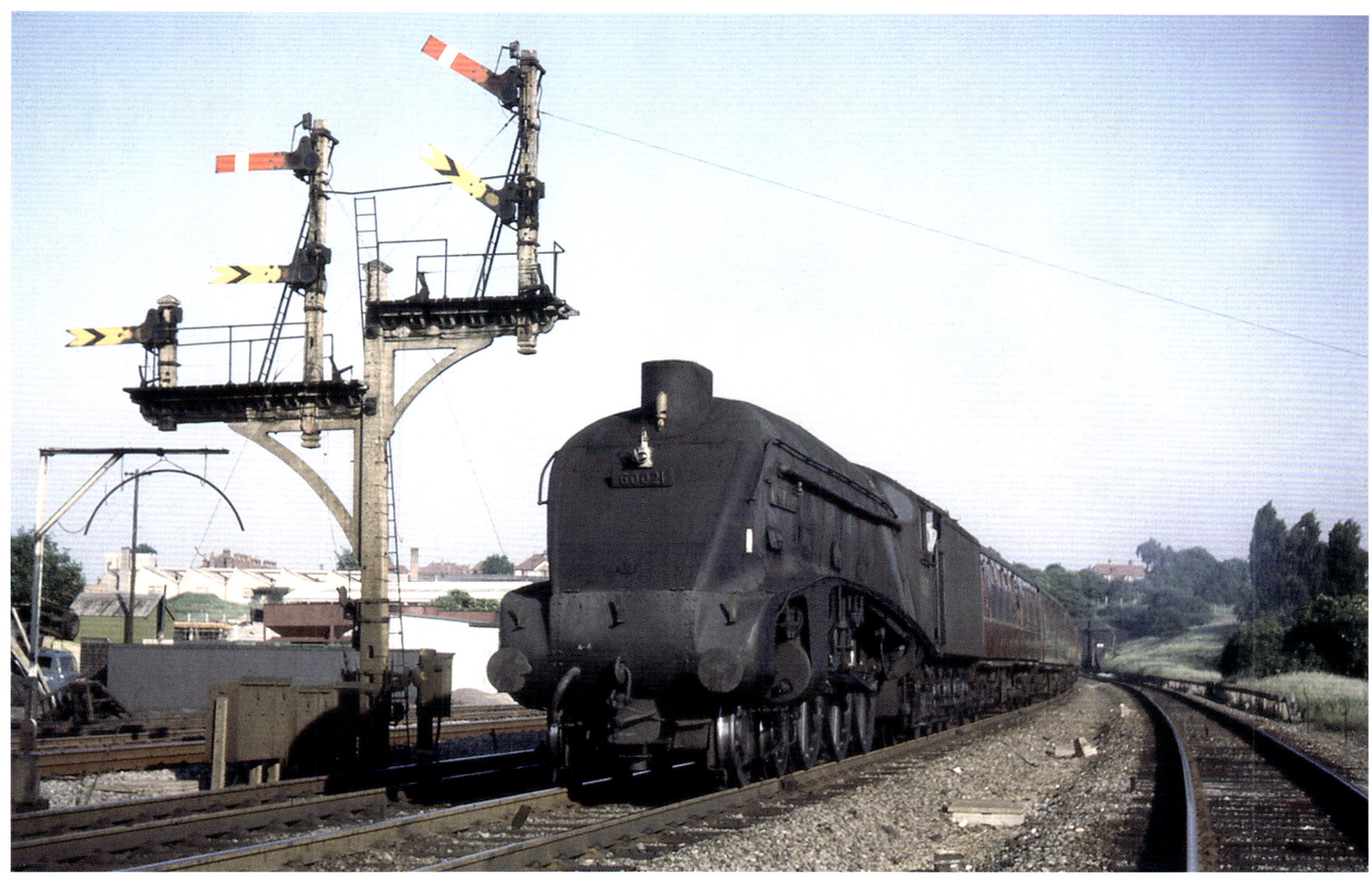

Opposite top: Sunday 9 June 1963. This day saw the running of the Home Counties Railway Society 'Doncaster Special', commencing at Kings Cross and using an unusual motive power for the East Coast Main Line in the shape of a Princess Coronation Class locomotive. Seen here entering Oakleigh Park station in the 'down' direction is No. 46245 *City of London* at the head of her twelve-coach train. Constructed at Crewe Works during 1943, she would be withdrawn in 1964. (Andrew G. Forsyth)

Opposite bottom: Monday 10 June 1963. Ex-LNER Class A4 4-6-2 No. 60021 *Wild Swan* is seen at Kings Cross station waiting at the head of the 6.26 p.m. departure to York. (Hugh D. Ramsay)

Above: Tuesday 11 June 1963. Leaning into the curve as it approaches New Southgate station, we see again No. 60021 *Wild Swan* at the head of the 6.17 p.m. relief working to Peterborough. Bearing a 34A Kings Cross shed code, she had been constructed at Doncaster Works in 1938 and numbered 4467 by the LNER. Allocated variously to Kings Cross, Grantham, Doncaster and finally New England shed, she was withdrawn four months after this photograph was taken, during October 1963. The impressive array of signals show both the New Southgate 'up' main starter and the Wood Green 'up' outer distant in the 'off' position. (Hugh D. Ramsay)

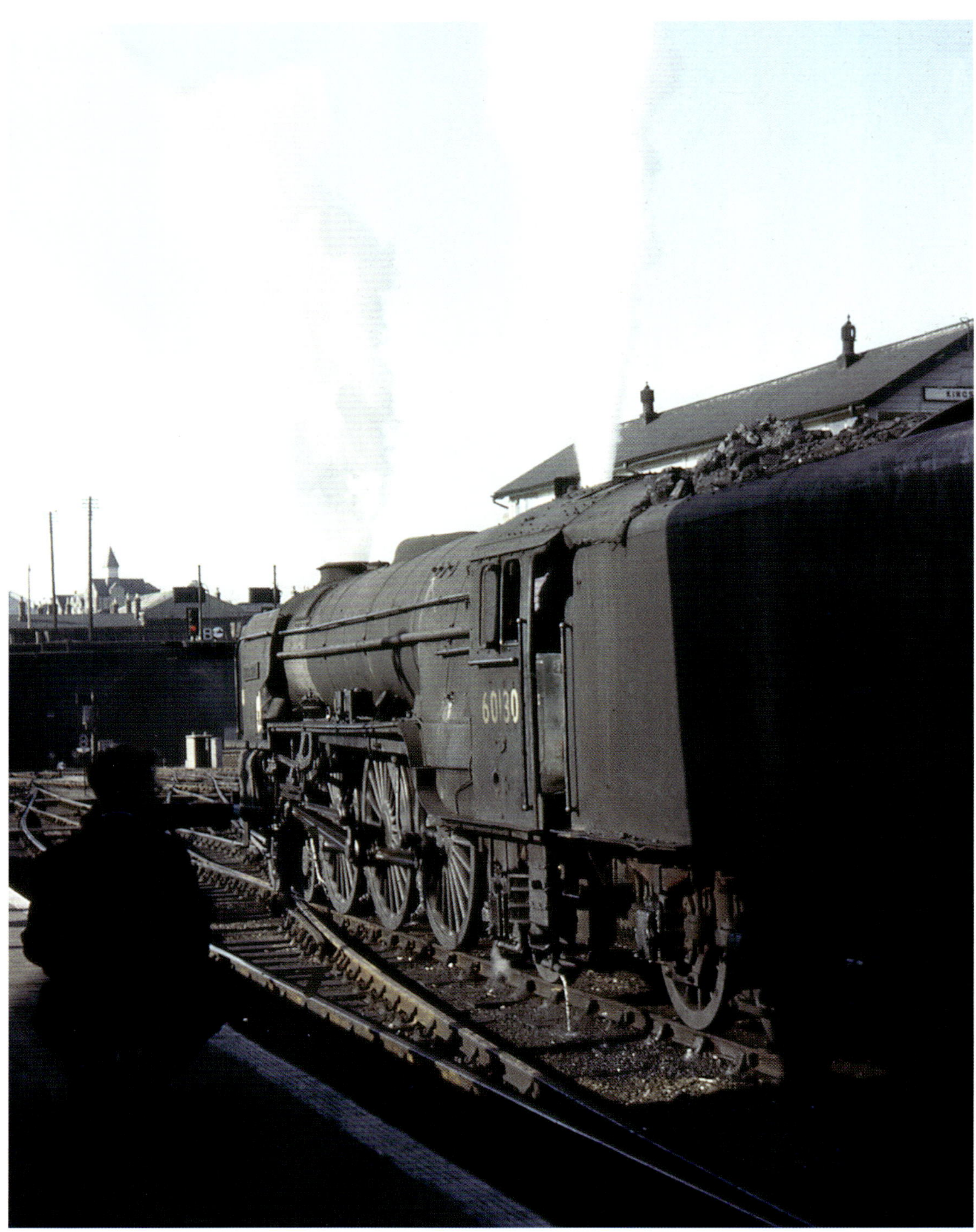

Left: Wednesday 12 June 1963. At Kings Cross station, in filthy condition, Class A1 4-6-2 No. 60130 *Kestrel* is on the point of departure from platform six at the head of the 6.12 p.m. departure to Leeds and is seen with both safety valves roaring. The first of the Darlington Works-constructed members of the class to enter service during 1948, she was allocated new to Doncaster and would spend time based at Kings Cross and Grantham, ending her days working out of Ardsley, being withdrawn in 1965. (Hugh D. Ramsay)

Opposite top: Monday 15 July 1963. On the approach to Seaton, a long train of goods wagons is being hauled by ex-LMS Class 8F 2-8-0 No. 48361. Entering service from Horwich Works during 1944, she would be withdrawn in 1966. (Andrew G. Forsyth)

Opposite bottom: Tuesday 16 July 1963. BR Standard Class 9F 2-10-0 No. 92094 is seen at the head of a train of empty 16-ton mineral wagons approaching Rugby on the Great Central Line. Constructed at Swindon Works in 1957, she would only give eleven years of service before being withdrawn during 1958. (Andrew G. Forsyth)

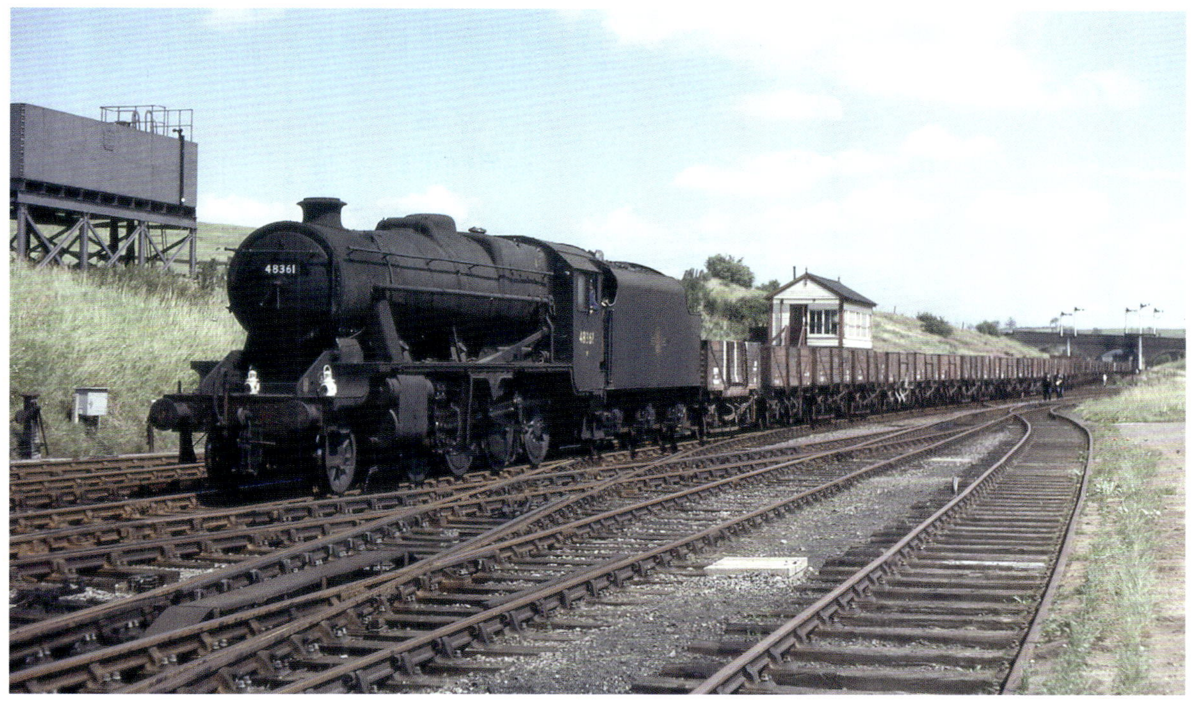

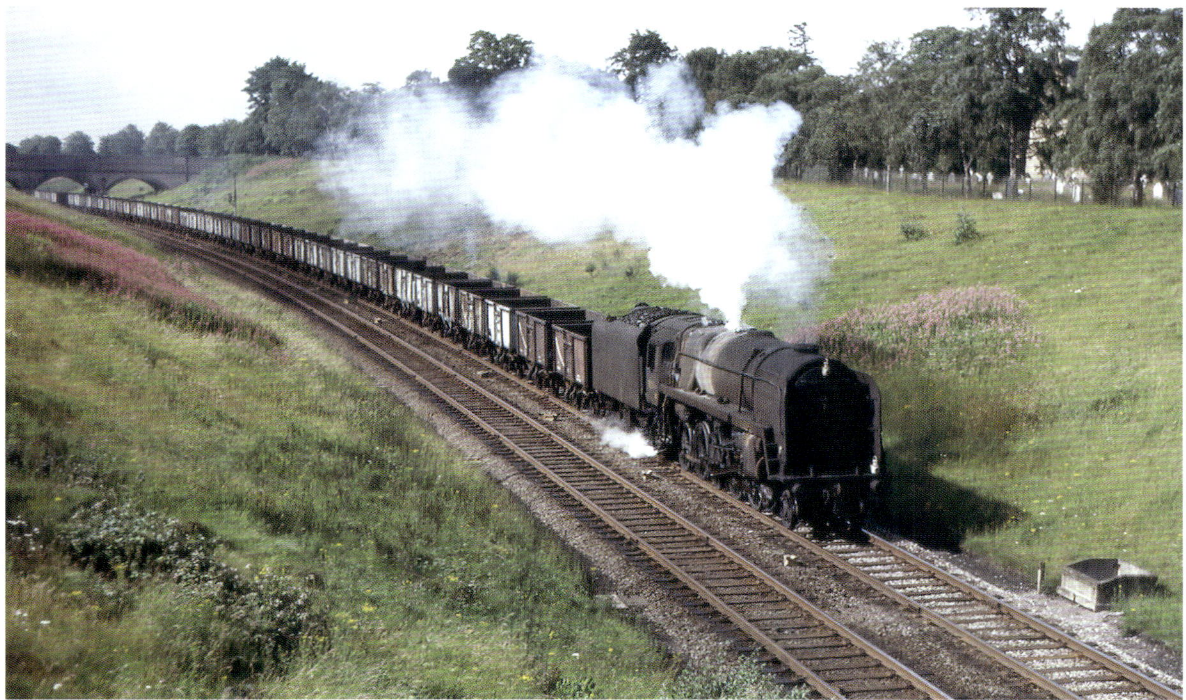

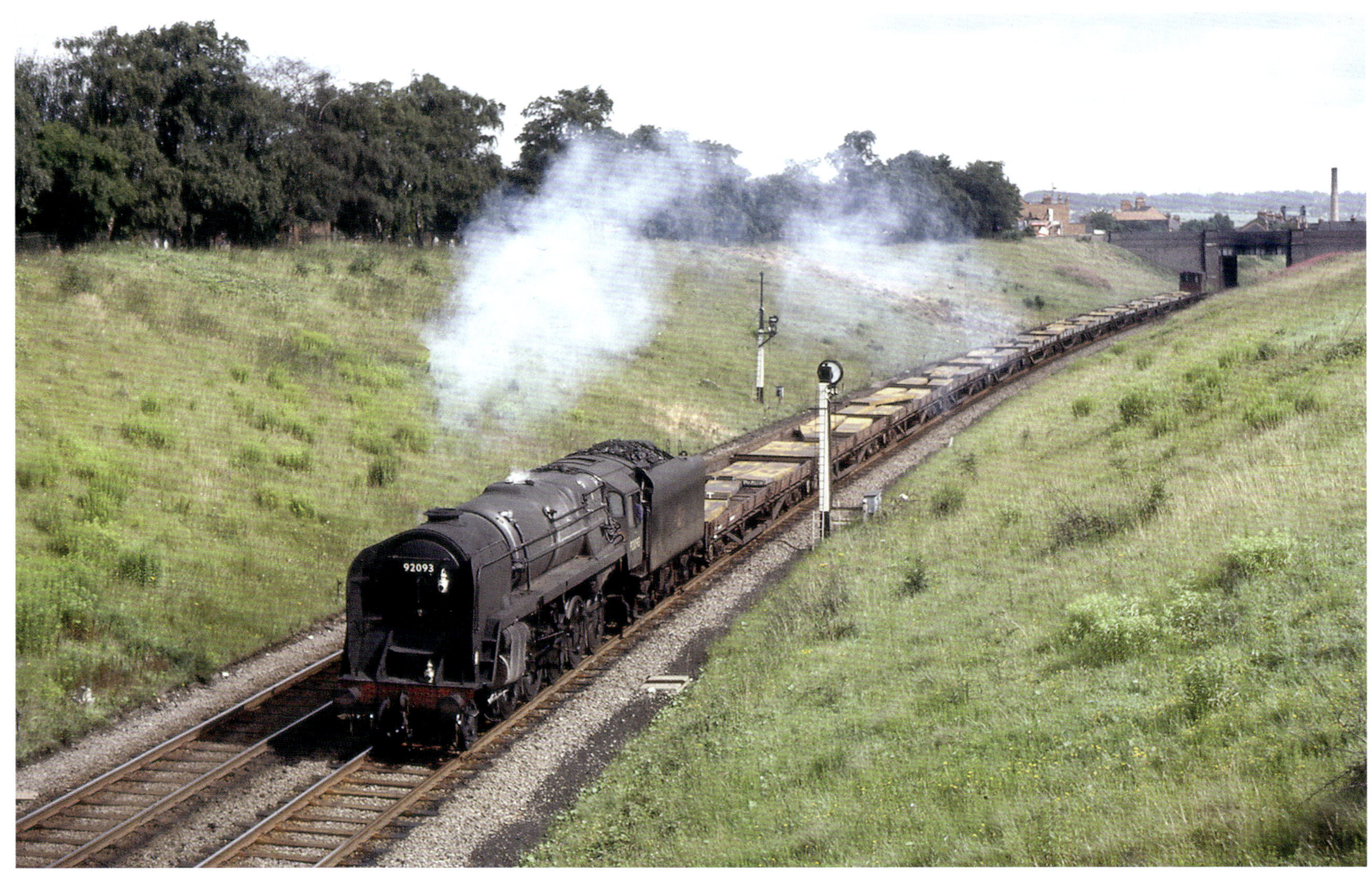

Above: Tuesday 16 July 1963. Seen near Rugby on the Great Central Main Line at the head of a train of loaded Boplate wagons is BR Standard Class 9F 2-10-0 No. 92093. Entering service from Swindon Works during 1957, she would have a short working life of only ten years, being withdrawn in 1967. (Andrew G. Forsyth)

Opposite top: Tuesday 16 July 1963. With the AEI Factory, the former British Thomson-Houston Works at Rugby, in the background, ex-LMS Class 5 2-6-0 No. 42958 is at the head of a 'down' goods train. Constructed at Crewe Works in 1934, she would be withdrawn from service during 1965. (Andrew G. Forsyth)

Opposite bottom: Saturday 20 July 1963. Seen at Stirling station in filthy condition is 65J Stirling-allocated ex-LMS Class 5 4-6-0 No. 45400 at the head of a Callander–Glasgow stopper. She had been the product of Armstrong Whitworth & Co. during 1937 and would end her days based at Stirling from 1950, being withdrawn in 1964. (Andrew G. Forsyth)

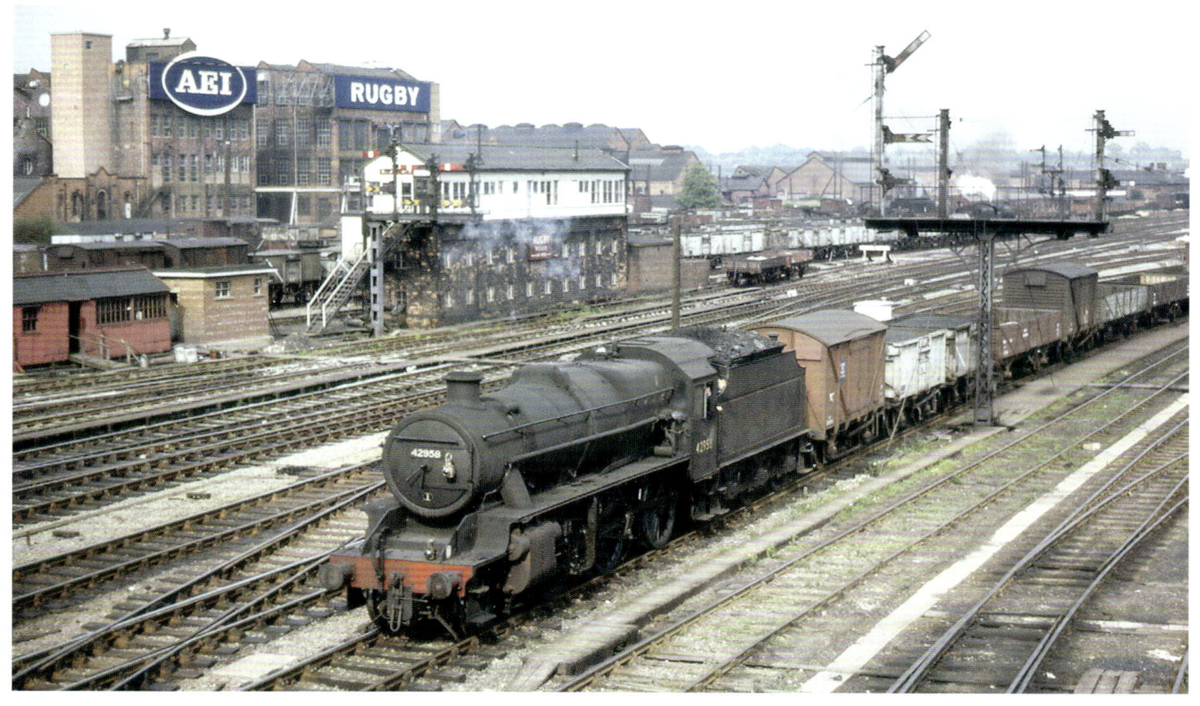
AEI RUGBY

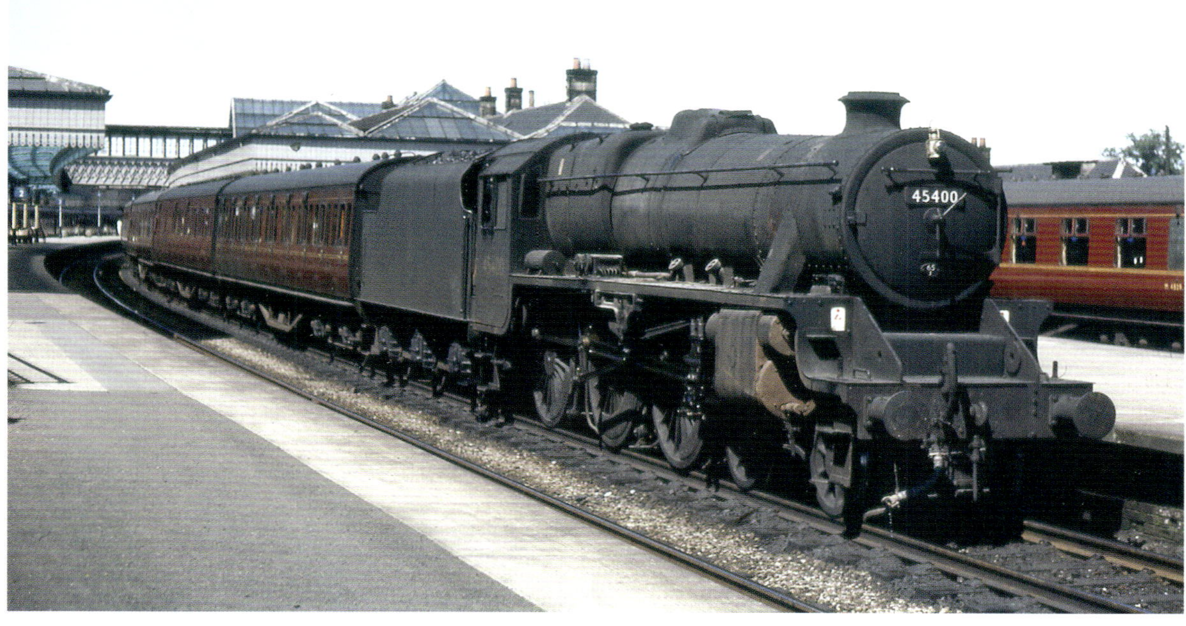
45400

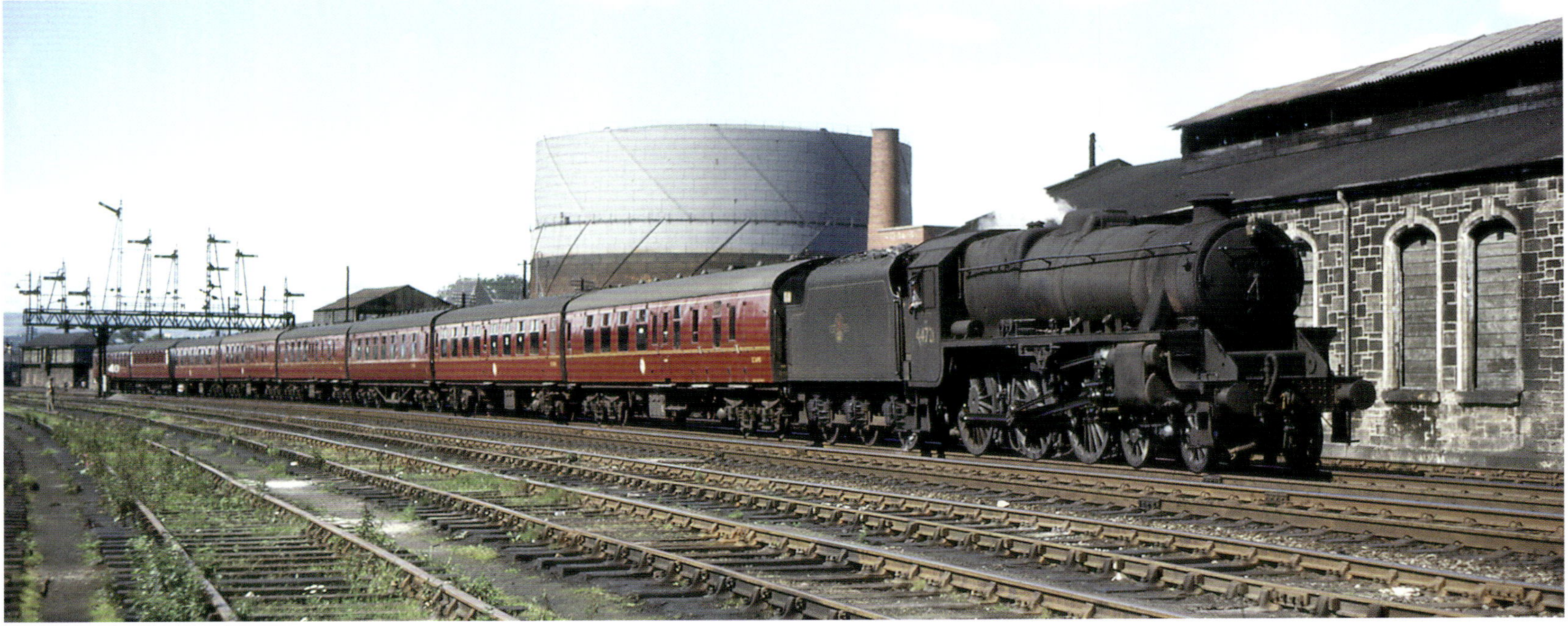

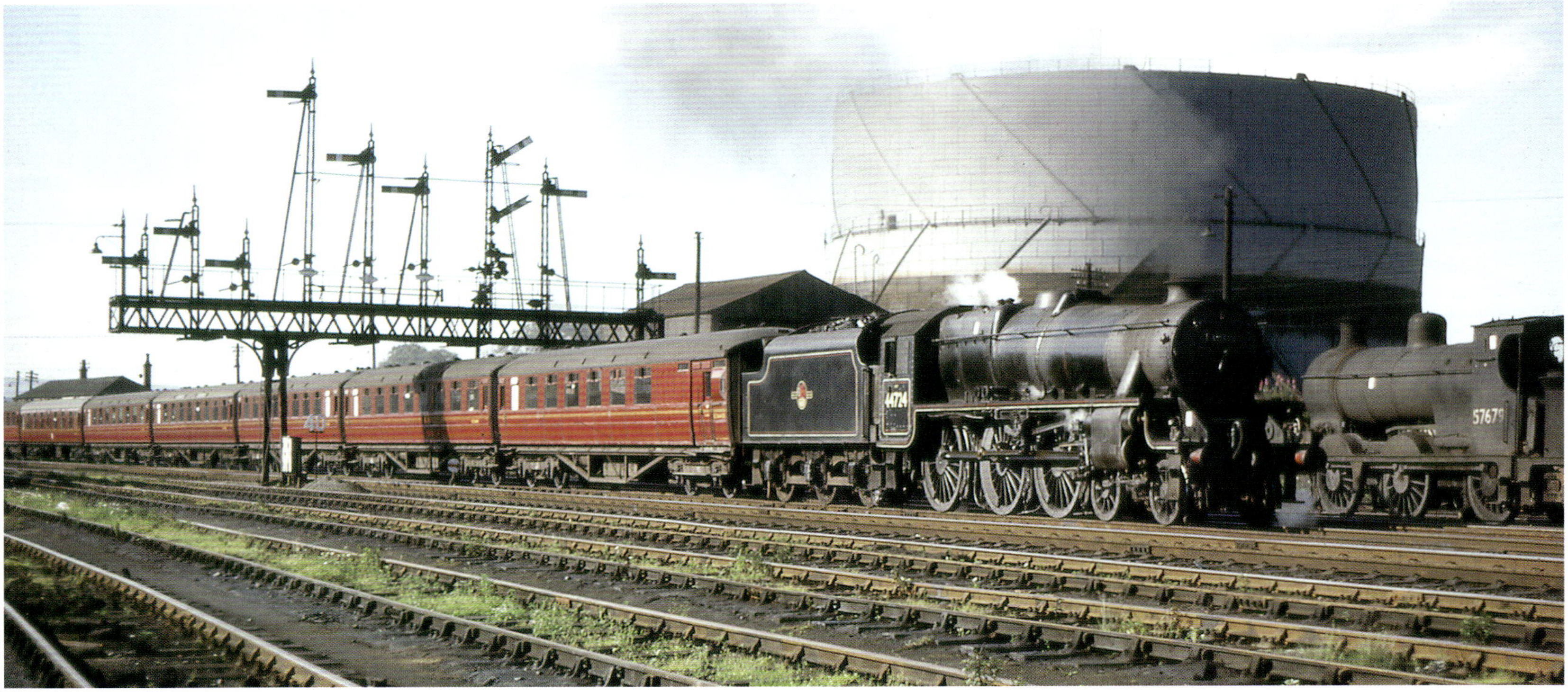

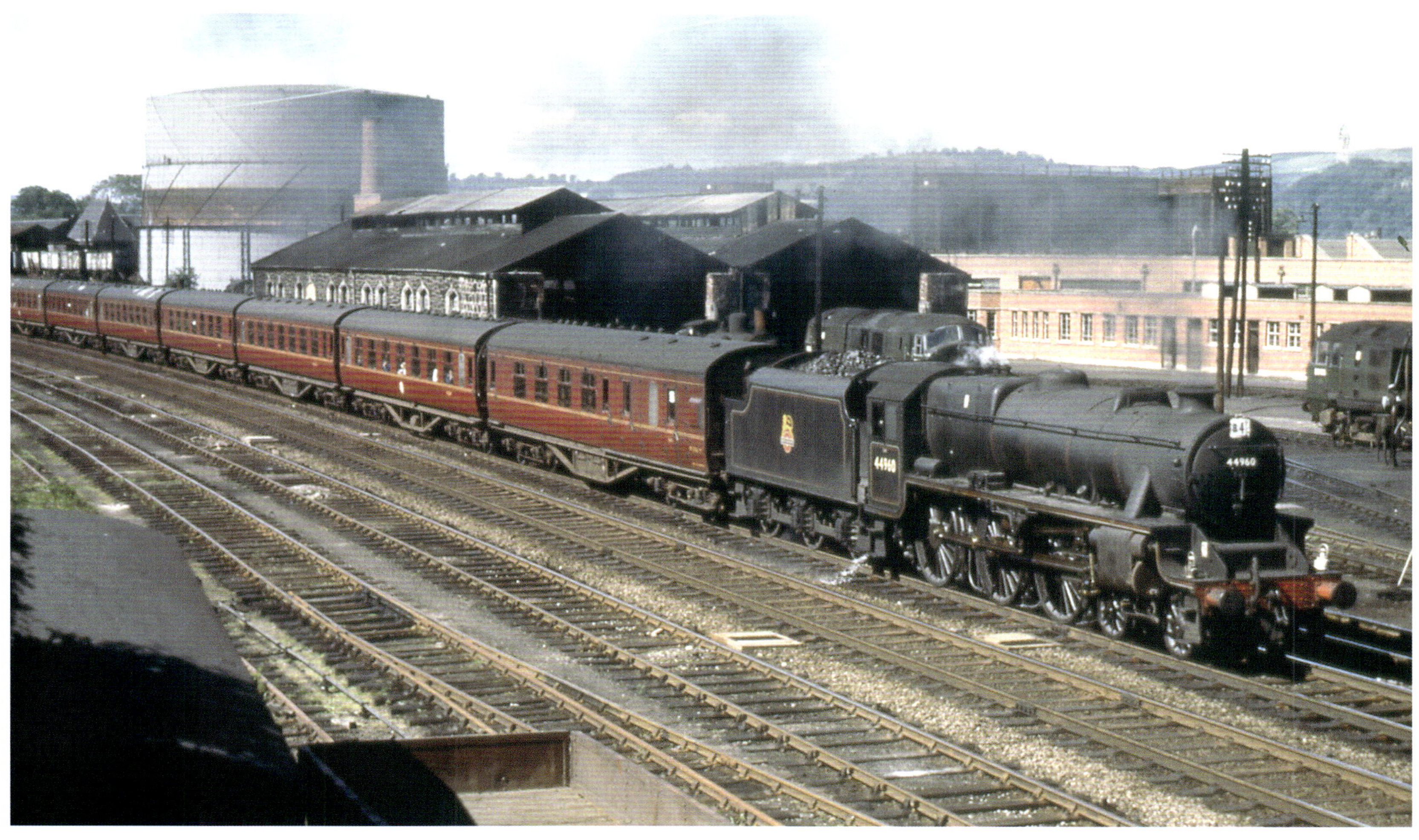

Saturday 20 July 1963. A visit to Stirling allowed the photographer to witness a constant stream of passenger traffic from Inverness, Aberdeen and Dundee to Glasgow and Edinburgh in both the 'up' and 'down' directions. This spread shows three of the William Stanier Black 5s in action.

Opposite top: Class 5 No. 44721 is seen about to pass Stirling shed at the head of an 'up' stopper. Constructed at Crewe Works during 1949, she would be withdrawn in 1965. (Andrew G. Forsyth)

Opposite bottom: Class 5 No. 44724, passing under the large south signal gantry, is at the head of a Dundee–Glasgow working. Entering service from Crewe Works in 1949, she would be withdrawn during 1966. (Andrew G. Forsyth)

Above: Ex-LMS Class 5 No. 44960 is passing the shed while working an Aberdeen–Glasgow express. A product of Horwich Works in 1946, she would be withdrawn twenty years later, during 1966. This highly successful class of locomotive extended to 842 members constructed between 1934 and 1951. Nos 44721 and 44724, seen opposite, were examples of only ten locomotives of the class fitted with steel fireboxes. (Andrew G. Forsyth)

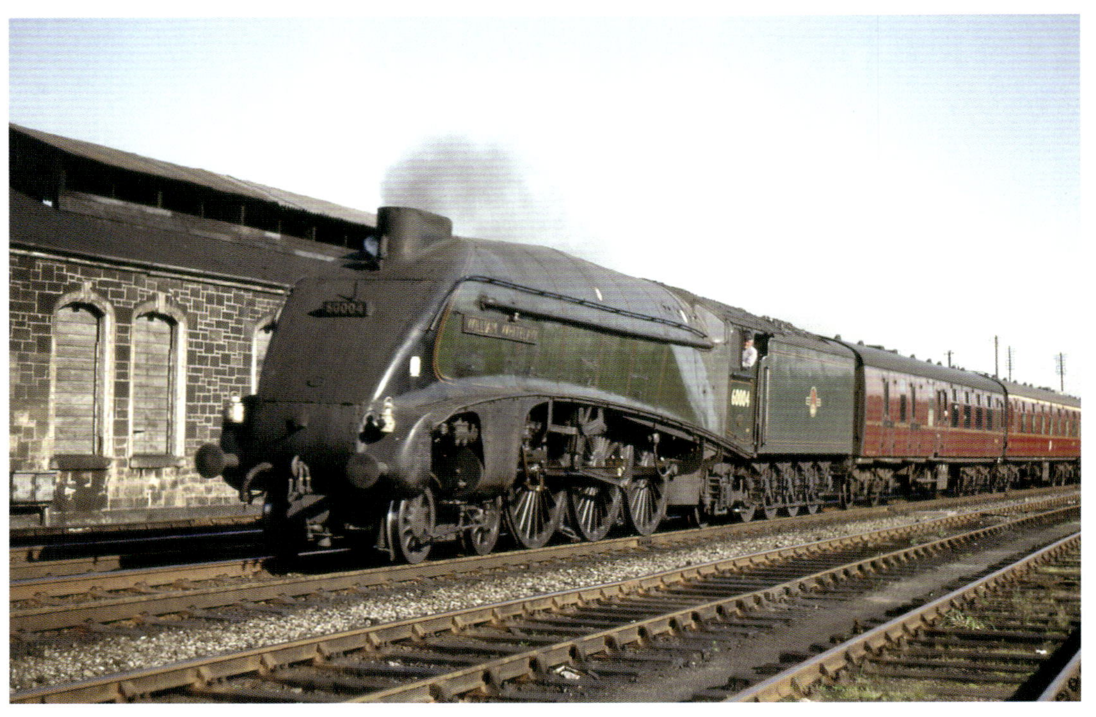

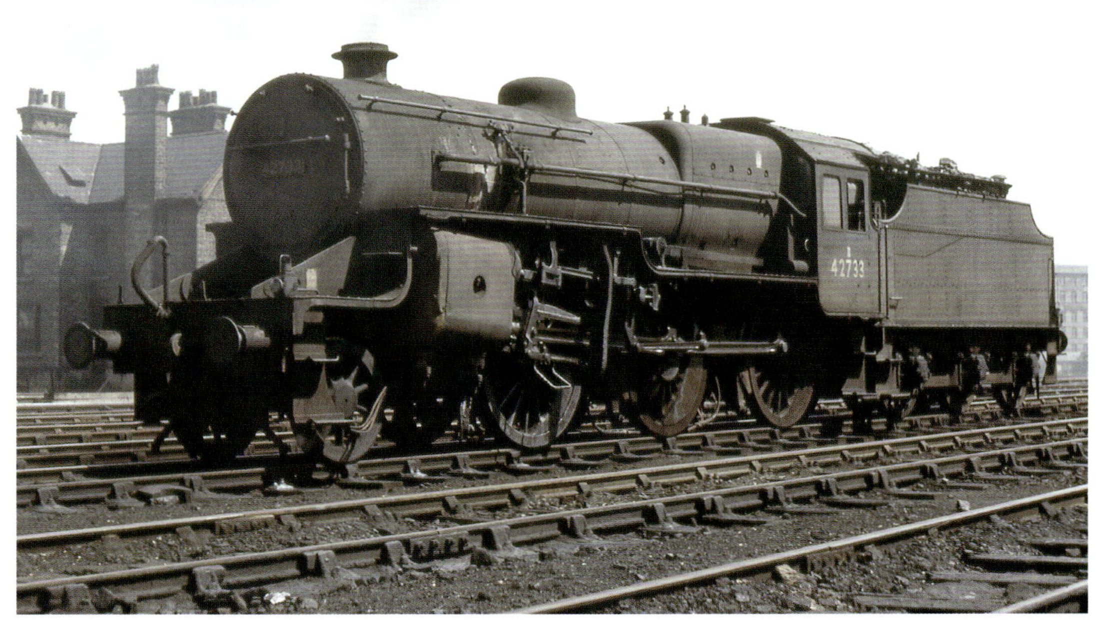

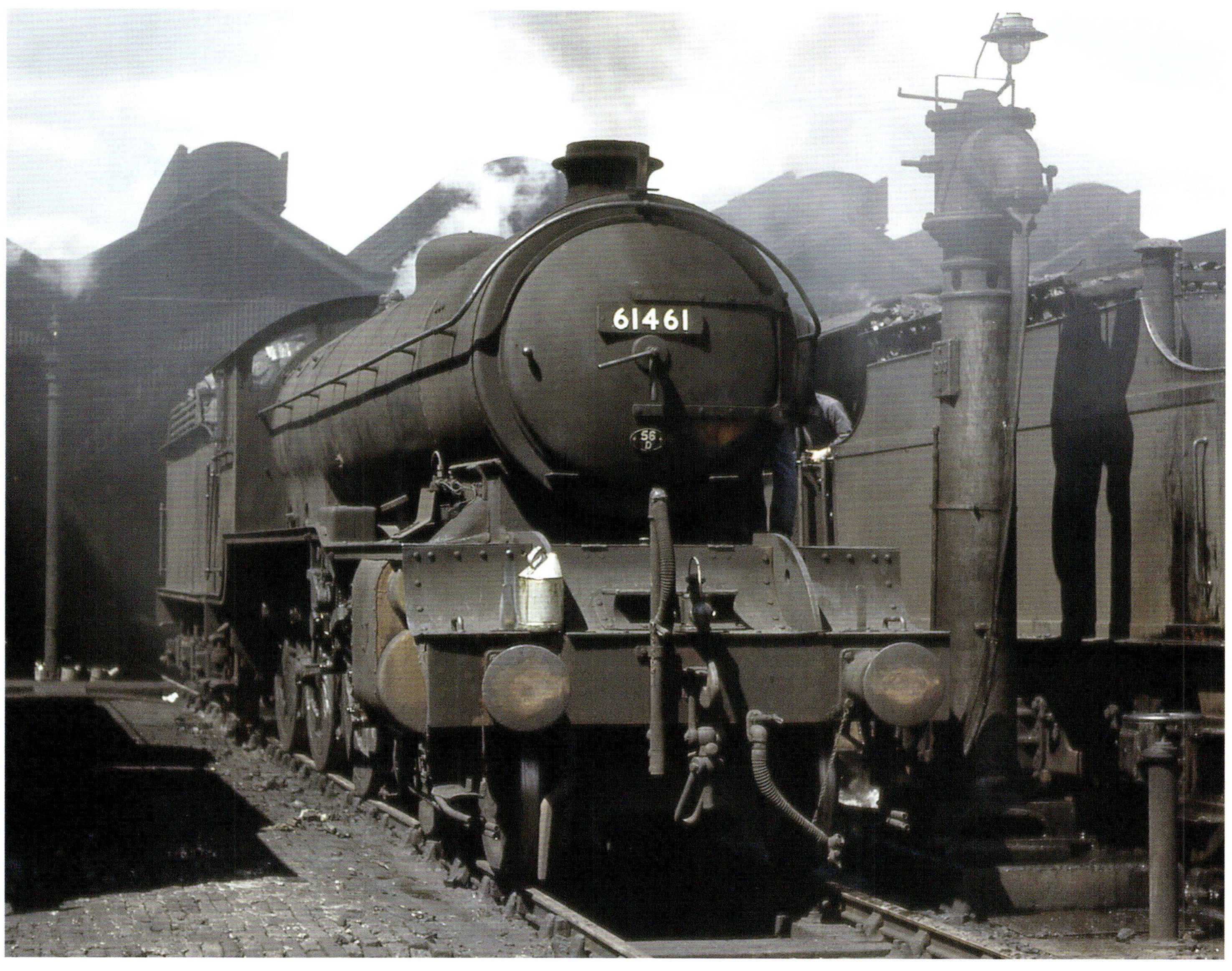

Opposite top: Saturday 20 July 1963. Ex-LNER Class A4 4-6-2 No. 60004 *William Whitelaw* is seen passing Stirling shed at the head of the 'down' 'St Mungo'. Only recently allocated to Ferryhill shed in Aberdeen, she would be kept in a good, clean condition by the staff there. (Andrew G. Forsyth)

Opposite bottom: Thursday 25 July 1963. Standing at Mirfield shed is ex-LMS Class 5 2-6-0 No. 42733. Constructed at Crewe Works during 1926, she would be withdrawn from service in 1965. (Andrew G. Forsyth)

Above: Thursday 25 July 1963. Also seen at Mirfield shed and bearing the correct 56D shed code is ex-LNER Class B16 4-6-0 No. 61461. This example of the Vincent Raven-designed class of three-cylinder locomotive had been constructed at Darlington Works, post-grouping, in 1923, utilising Stephenson valve gear. She would be rebuilt during 1948 with Walschaerts valve gear, as seen here, and would be withdrawn from service two months after this photograph was taken, in September 1963. (Andrew G. Forsyth)

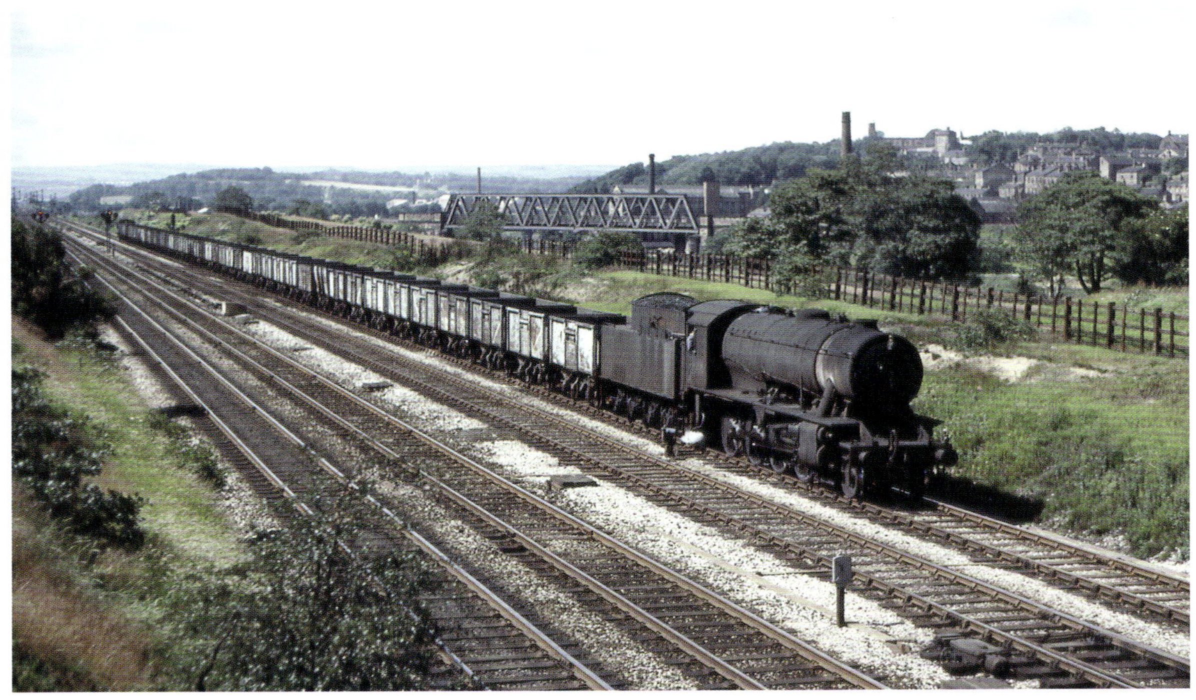

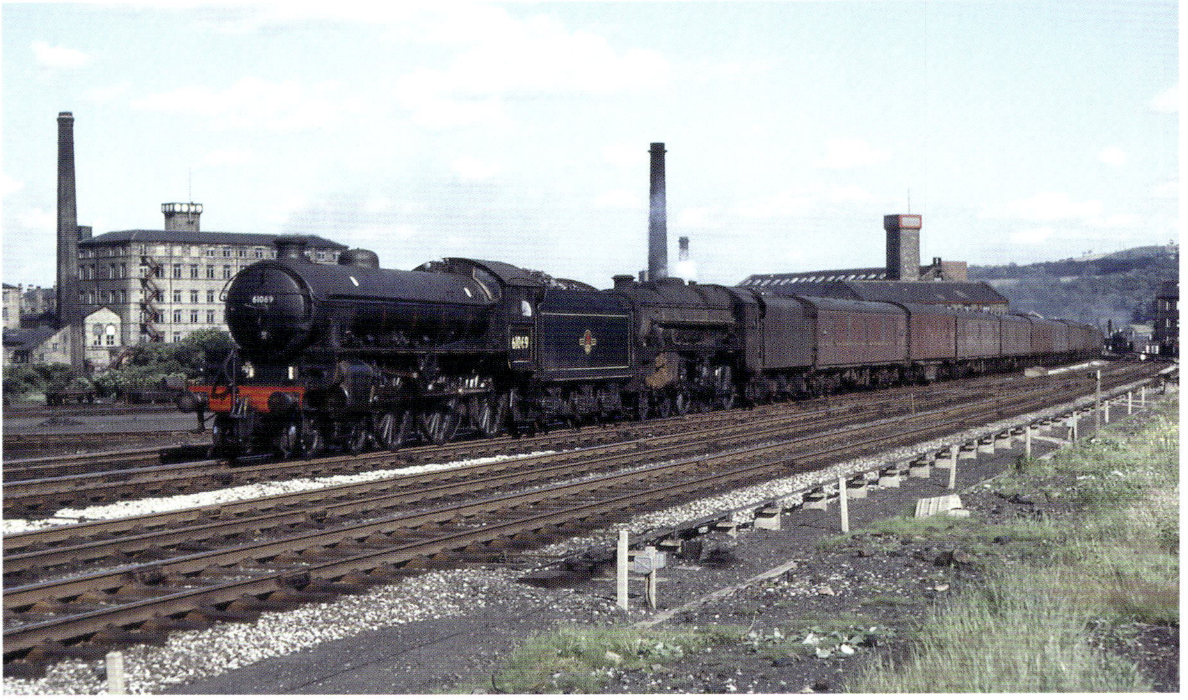

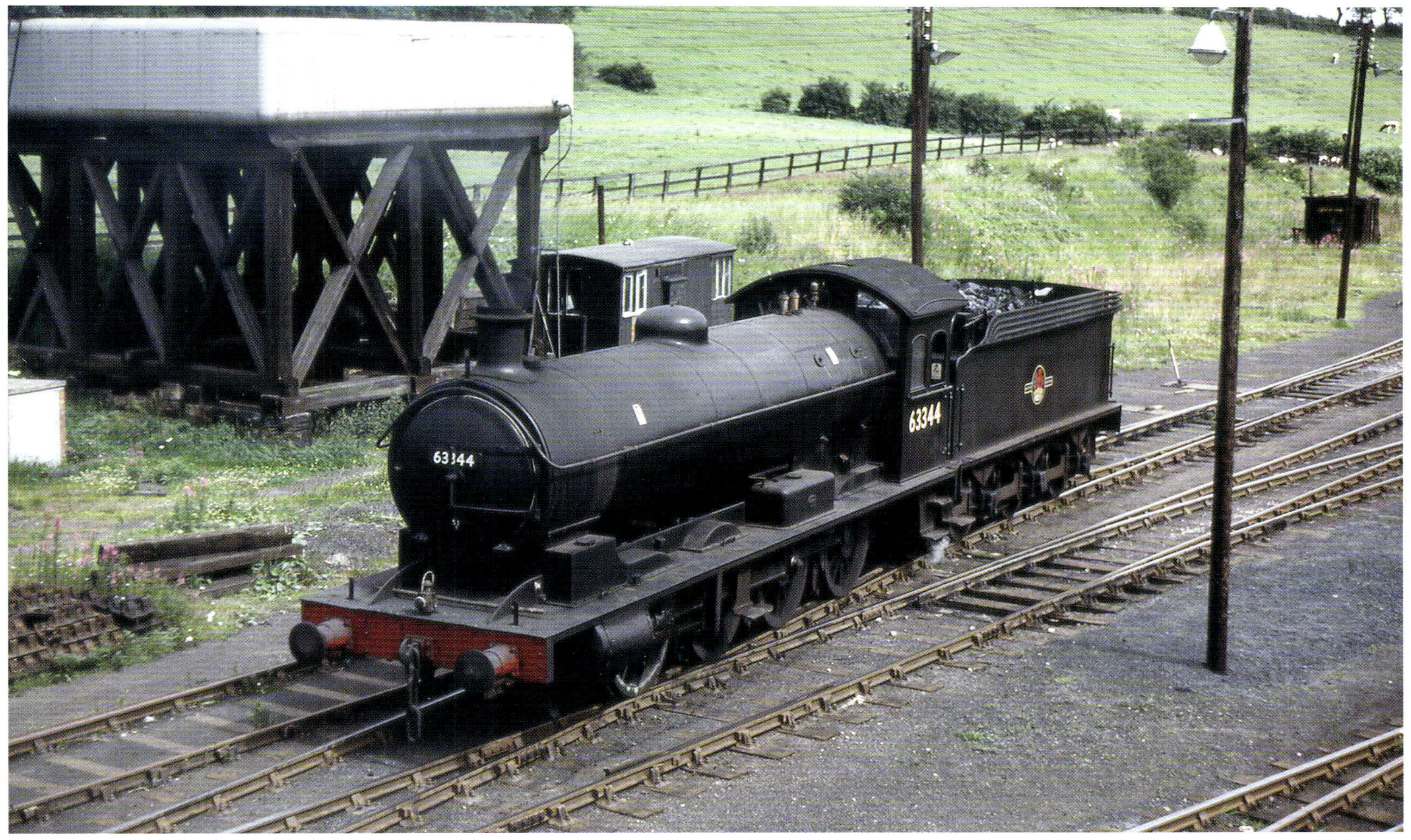

Opposite top: Thursday 25 July 1963. Approaching Mirfield with a long train of unfitted 16-ton mineral wagons is Ministry of Supply Austerity Class 8F 2-8-0 No. 90584. A product of the Vulcan Foundry in 1943 and numbered 77129 by the War Department, she would be purchased by British Railways in 1948 and withdrawn from service during 1964. (Andrew G. Forsyth)

Opposite bottom: Thursday 25 July 1963. Seen near Mirfield shed is a York–Manchester parcels working with an unusual pairing of motive power at its head. Ex-LNER Class B1 4-6-0 No. 61069 in clean condition is piloting Class 5 4-6-0 No. 44696. The B1 was a product of the NBL during 1946 that would be withdrawn the month following this photograph being taken, August 1963. The Black 5 had been constructed at Horwich Works in 1950 and would be withdrawn during 1967. (Andrew G. Forsyth)

Above: Friday 26 July 1963. At Shildon goods sidings, ex-NER Class T2 (LNER Class Q6) 0-8-0 No. 63344 is moving into position for its next duty. Constructed at Darlington Works during 1913, she would be withdrawn from service in 1964. (Andrew G. Forsyth)

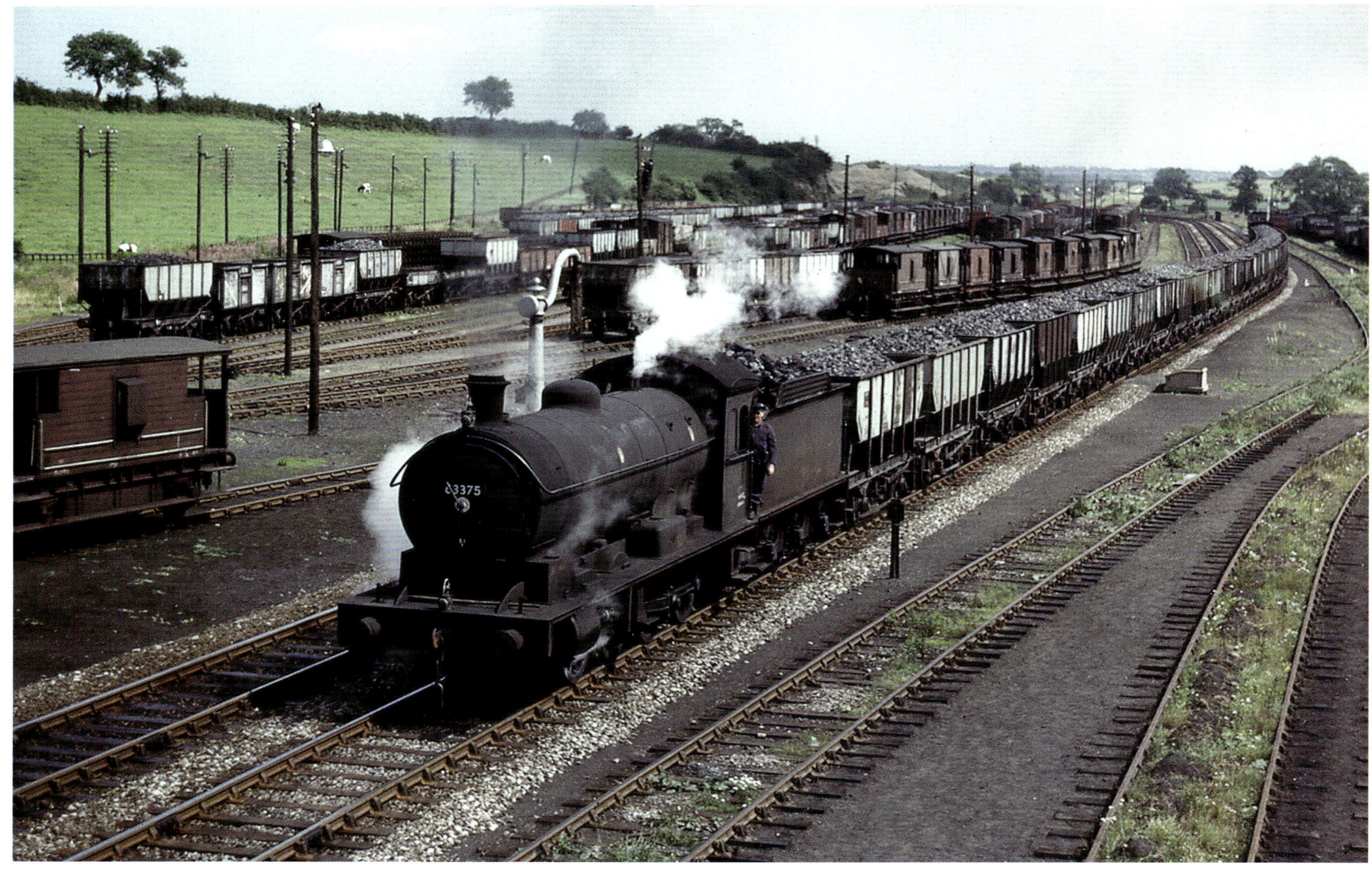

Above: Friday 26 July 1963. Seen passing Shildon goods sidings with a train of loaded mineral hopper wagons, ex-NER Class T2 (LNER Class Q6) 0-8-0 No. 63375 has steam to spare. A product of Darlington Works in 1917, she would be withdrawn during the month following the taking of this photograph, August 1963. (Andrew G. Forsyth)

Opposite top: Saturday 27 July 1963. Ex-LNER Class A4 4-6-2 No. 60023 *Golden Eagle* is seen departing from York station at the head of a 'down' express. Bearing a 52A Gateshead shed code, she had been constructed at Doncaster Works during 1936 and would end her days based at Ferryhill shed in Aberdeen, being withdrawn in 1964. (Andrew G. Forsyth)

Opposite bottom: Saturday 27 July 1963. Standing in York station at the head of a Scarborough-bound train is ex-LNER three-cylinder Class B16/3 4-6-0 No. 61454. Entering service from Darlington Works during 1923 utilising Stephenson valve gear, she would be rebuilt as seen here with Walschaerts valve gear in 1944, and was later withdrawn from service in 1964. (Andrew G. Forsyth)

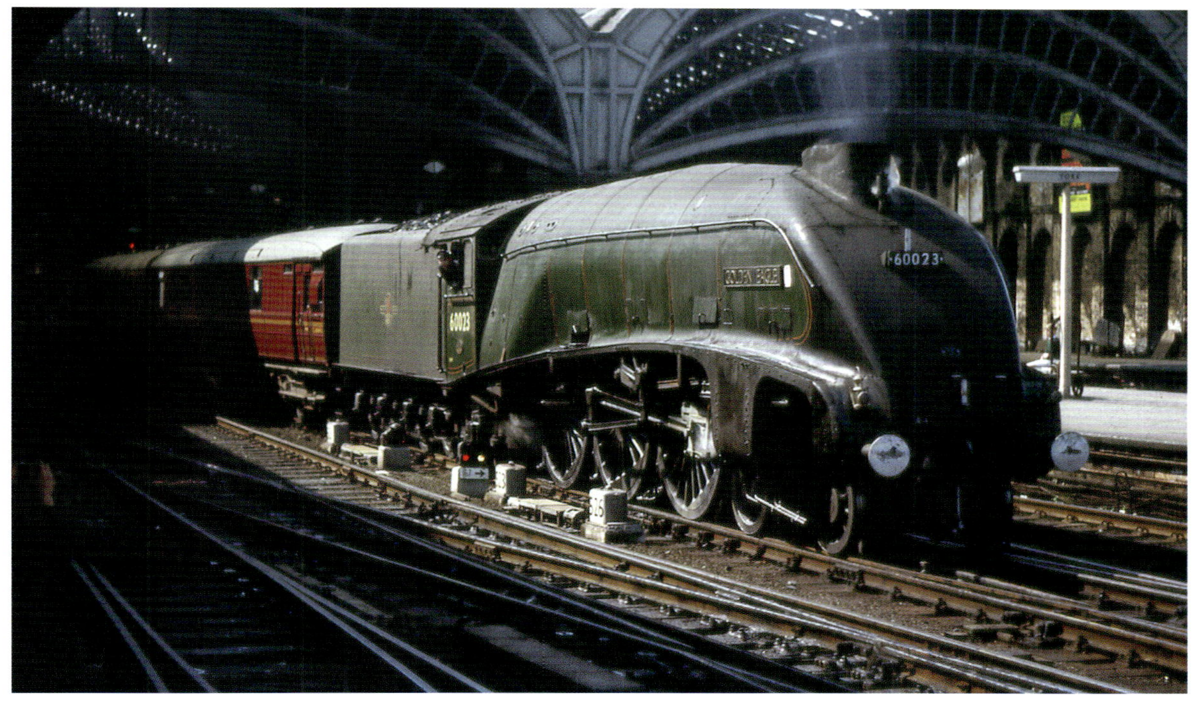

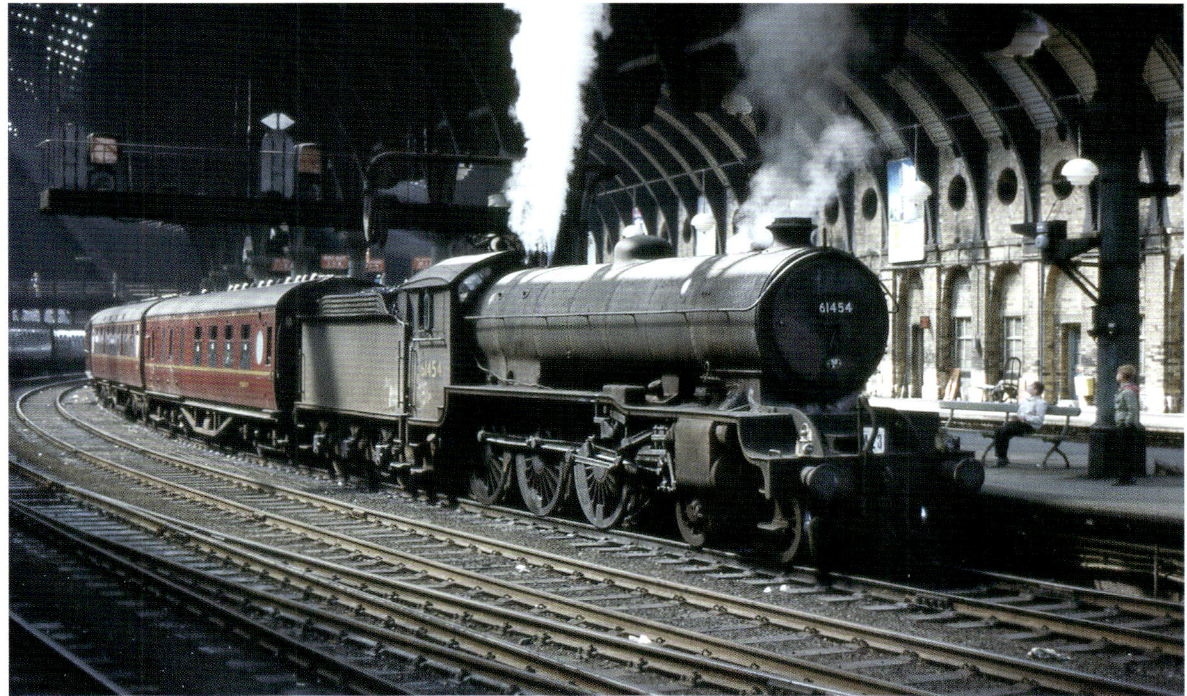

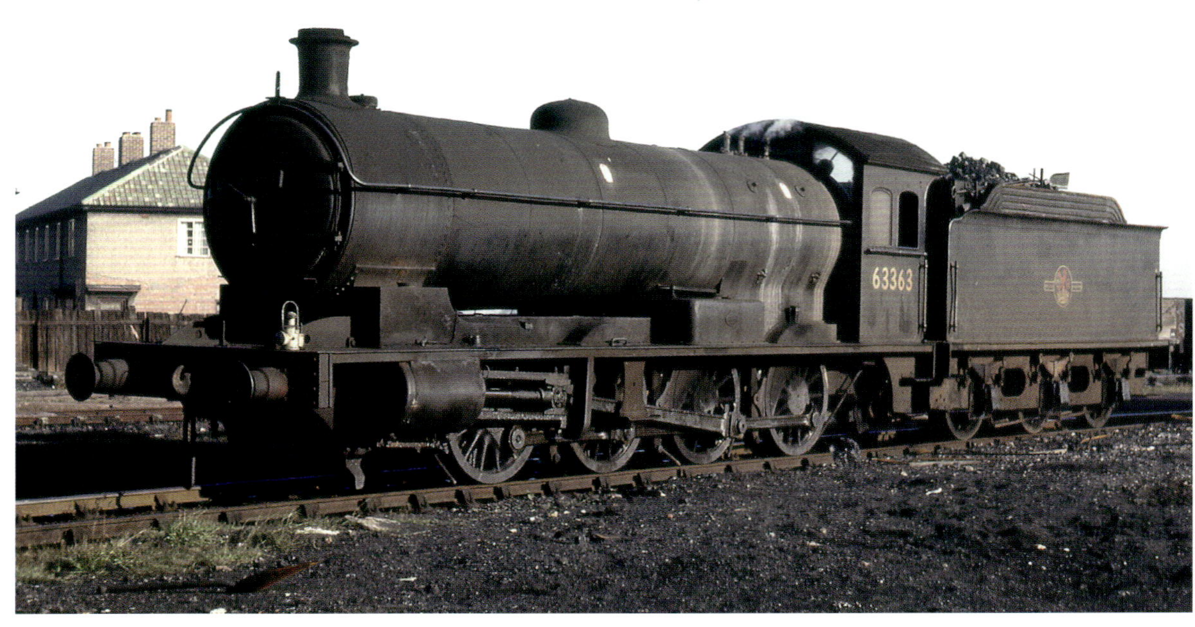

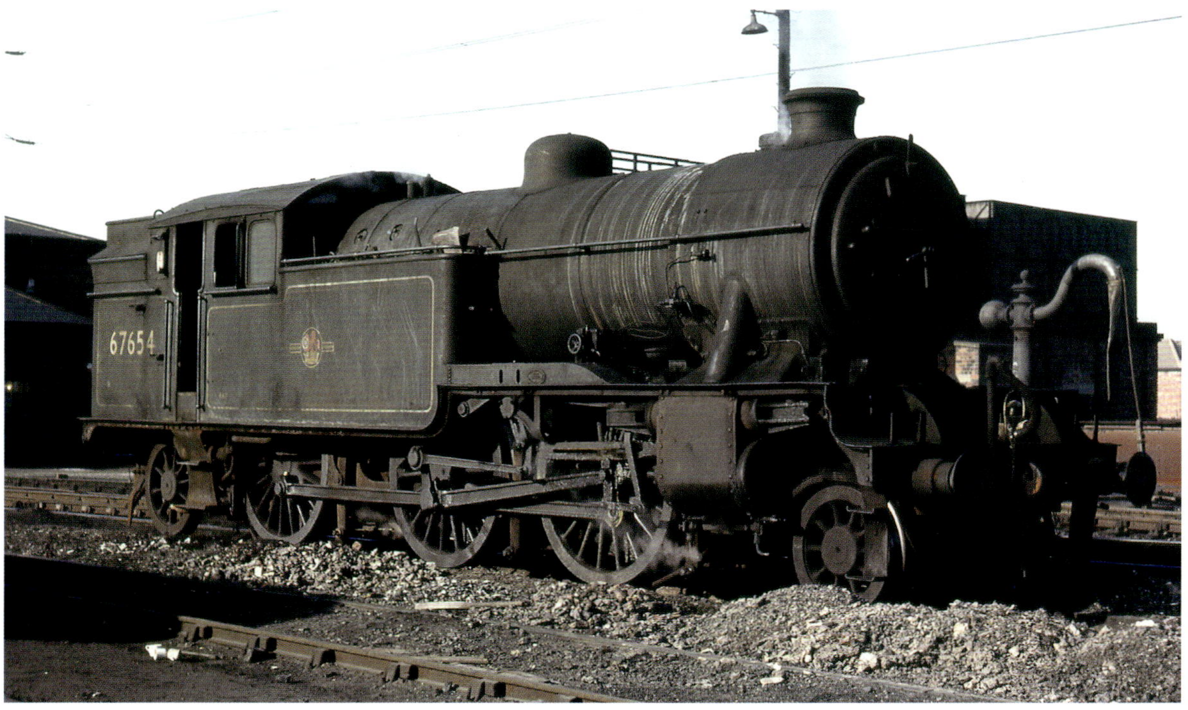

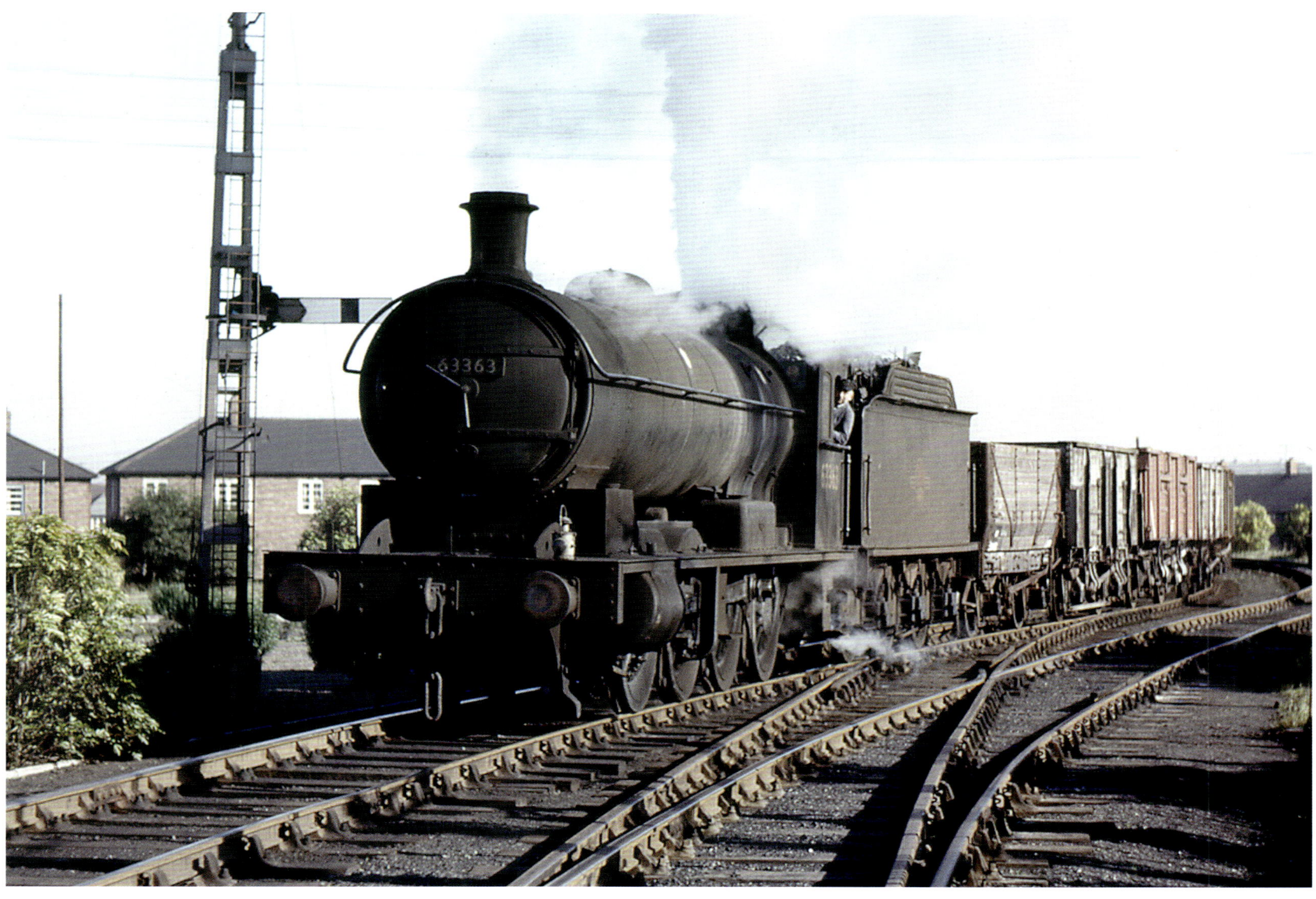

Opposite top: Saturday 28 September 1963. At Tyne Dock shed, ex-NER Class T2 (LNER Class Q6) 0-8-0 No. 63363 is standing in the yard. Constructed at Darlington Works during 1913 and allocated new to Tyne Dock shed, she would be withdrawn fifty-three years later, while based there, in 1966. (Andrew G. Forsyth)

Opposite bottom: Saturday 28 September 1963. Also seen in Tyne Dock shed yard is ex-LNER Class V3 2-6-2 tank No. 67654 in appalling external condition. Entering service from Doncaster Works in 1936 as a Class V1 locomotive, she would be rebuilt during 1954 as a Class V3 and withdrawn from service days after this photograph was taken. (Andrew G. Forsyth)

Above: Saturday 28 September 1963. Ex-NER Class T2 (LNER Class Q6) 0-8-0 No. 63363 is seen here again later the same day at the head of a train of mineral wagons passing Tyne Dock station. (Andrew G. Forsyth)

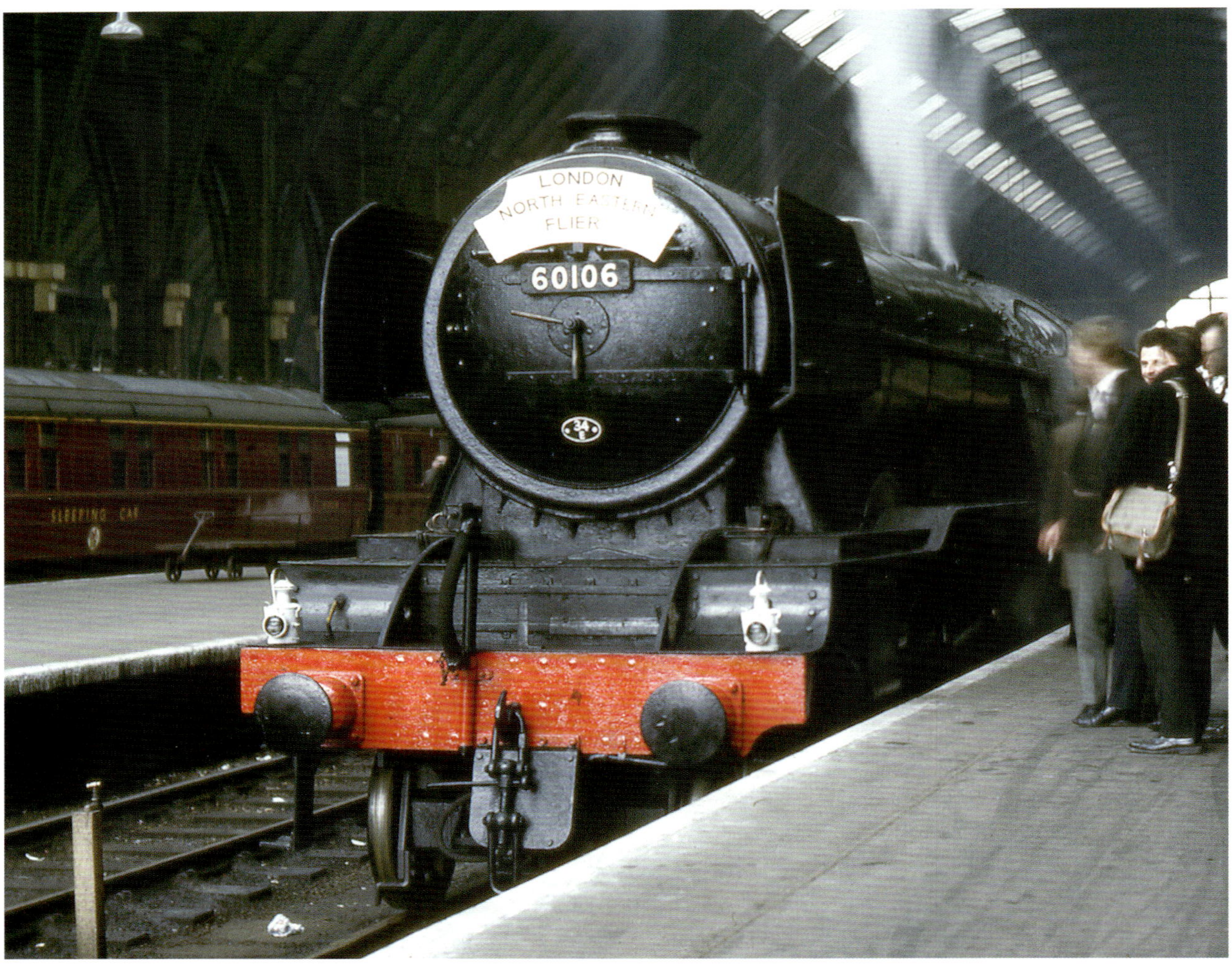

Above: Saturday 2 May 1964. In suitably clean condition, ex-LNER Class A3 4-6-2 No. 60106 *Flying Fox* is seen at Kings Cross station waiting to depart at the head of the Gresley Society-organised 'London North Eastern Flyer' to Doncaster. Constructed at Doncaster Works during 1923 as a Class A1 locomotive, she would be rebuilt in 1947 as a Class A3 locomotive, acquiring a double chimney in 1958 and trough-style smoke deflectors in 1961, and later be withdrawn from service at the end of 1964. Bearing a 34E New England shed code, she was named after the racehorse that had won the 1899 Derby, St Leger and 2,000 Guineas races. (Andrew G. Forsyth)

Opposite top: Saturday 2 May 1964. Waiting to depart from Euston station is ex-LMS Class 5 4-6-0 No. 45379. A product of Armstrong Whitworth & Co. during 1937, she is seen bearing a 1A Willesden shed code. Withdrawn from service in 1965, she lay in Barry scrapyard until 1974, when purchased for preservation. She is currently in store awaiting overhaul at the Mid-Hants Railway. (Andrew G. Forsyth)

Opposite bottom: Monday 8 June 1964. Approaching York station with a goods train in the northbound direction is Class 4MT 2-6-0 No. 43126, looking in relatively clean condition. Constructed at Horwich Works during 1951, she would be allocated to York shed in 1961 and withdrawn from service in 1966 while based there. In the background is St Paul's church in Holgate Street, which appears in many railway photographs taken in the area. (Andrew G. Forsyth)

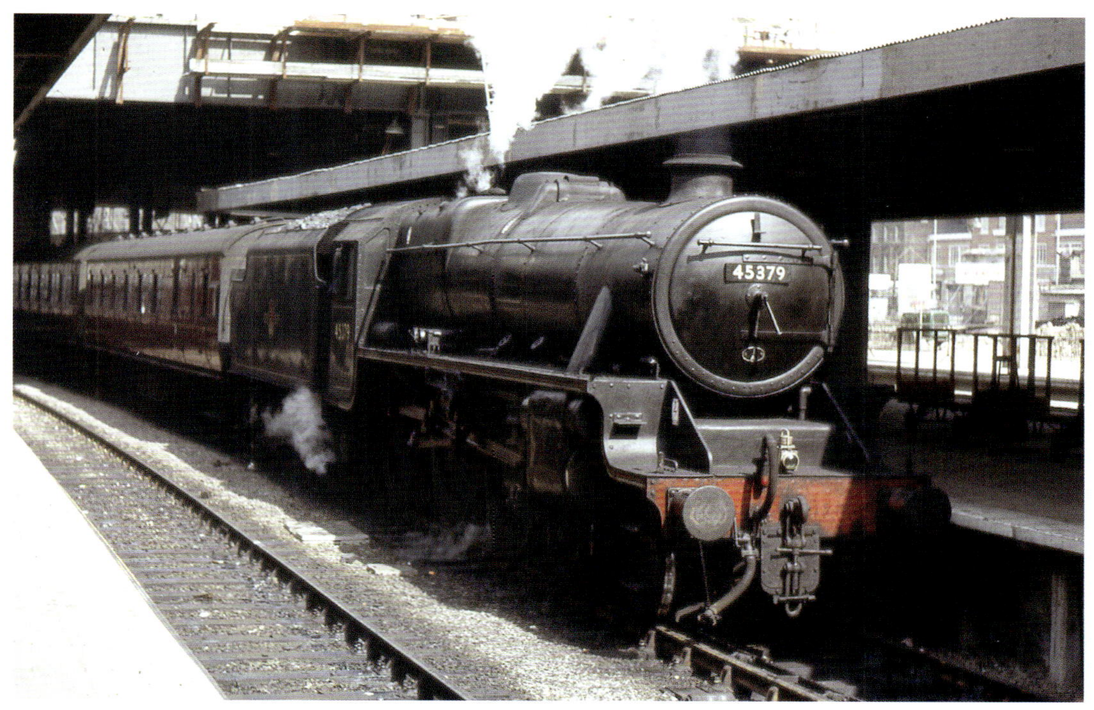

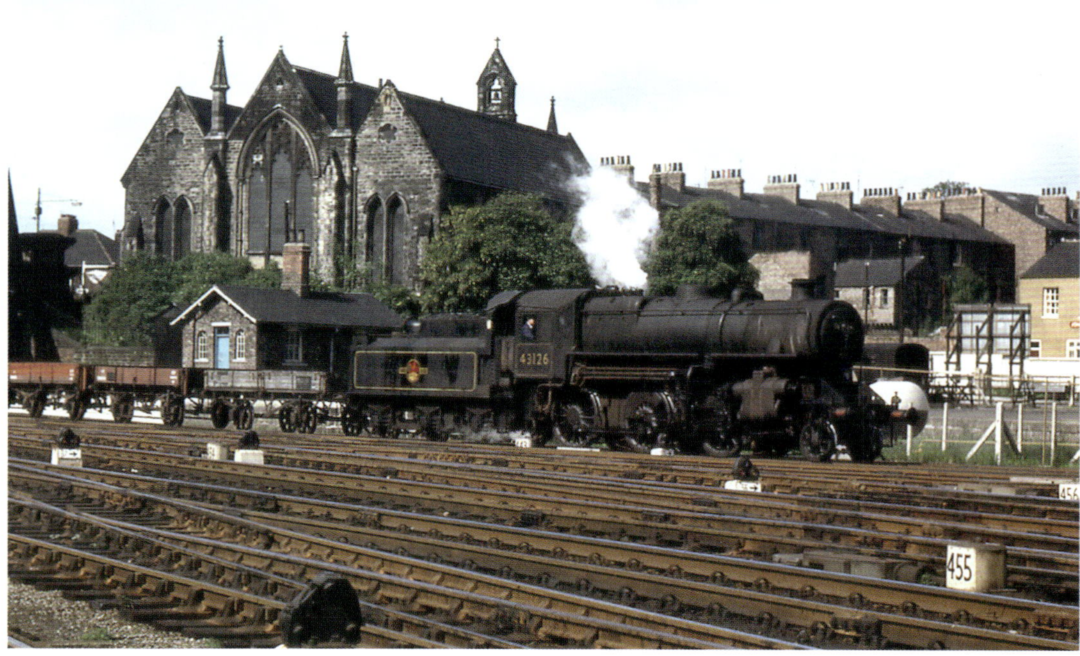

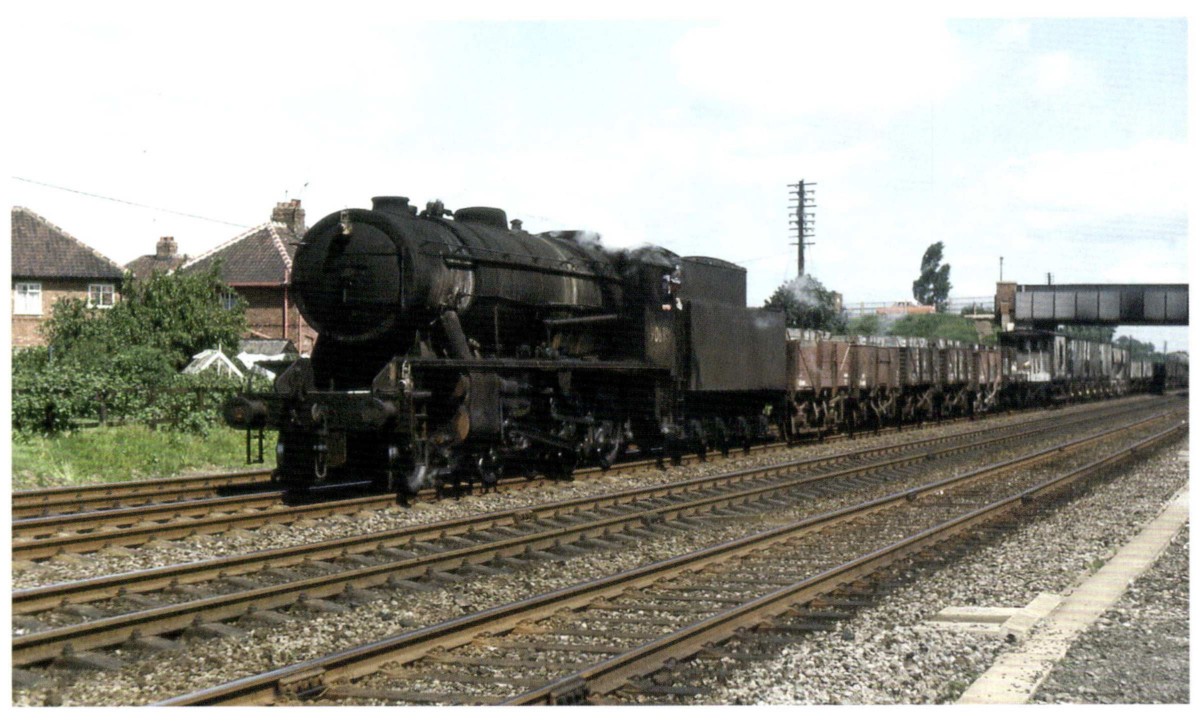

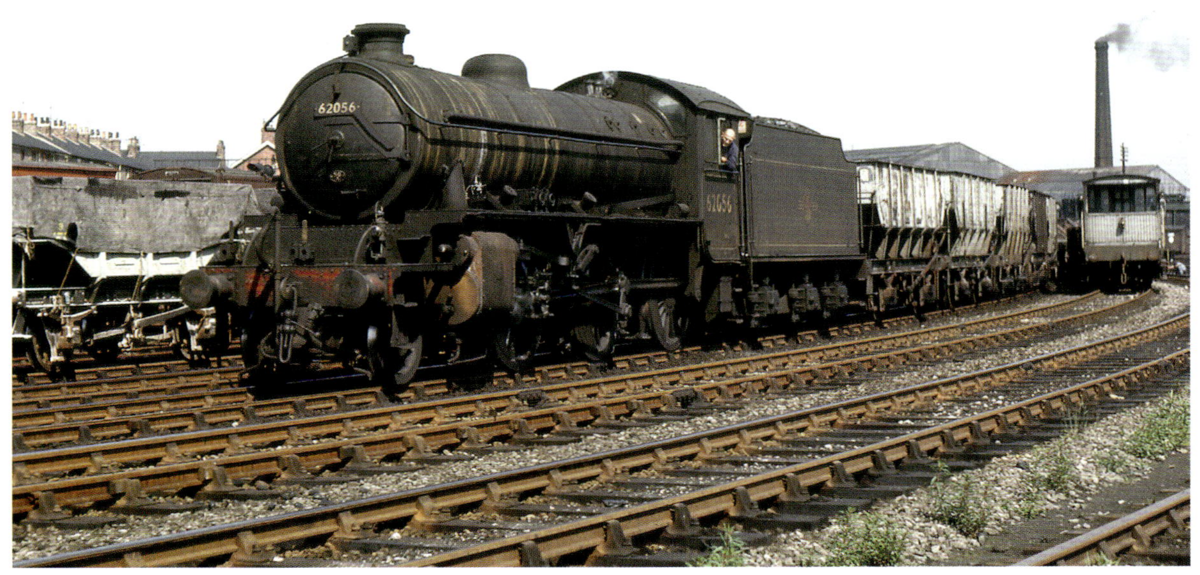

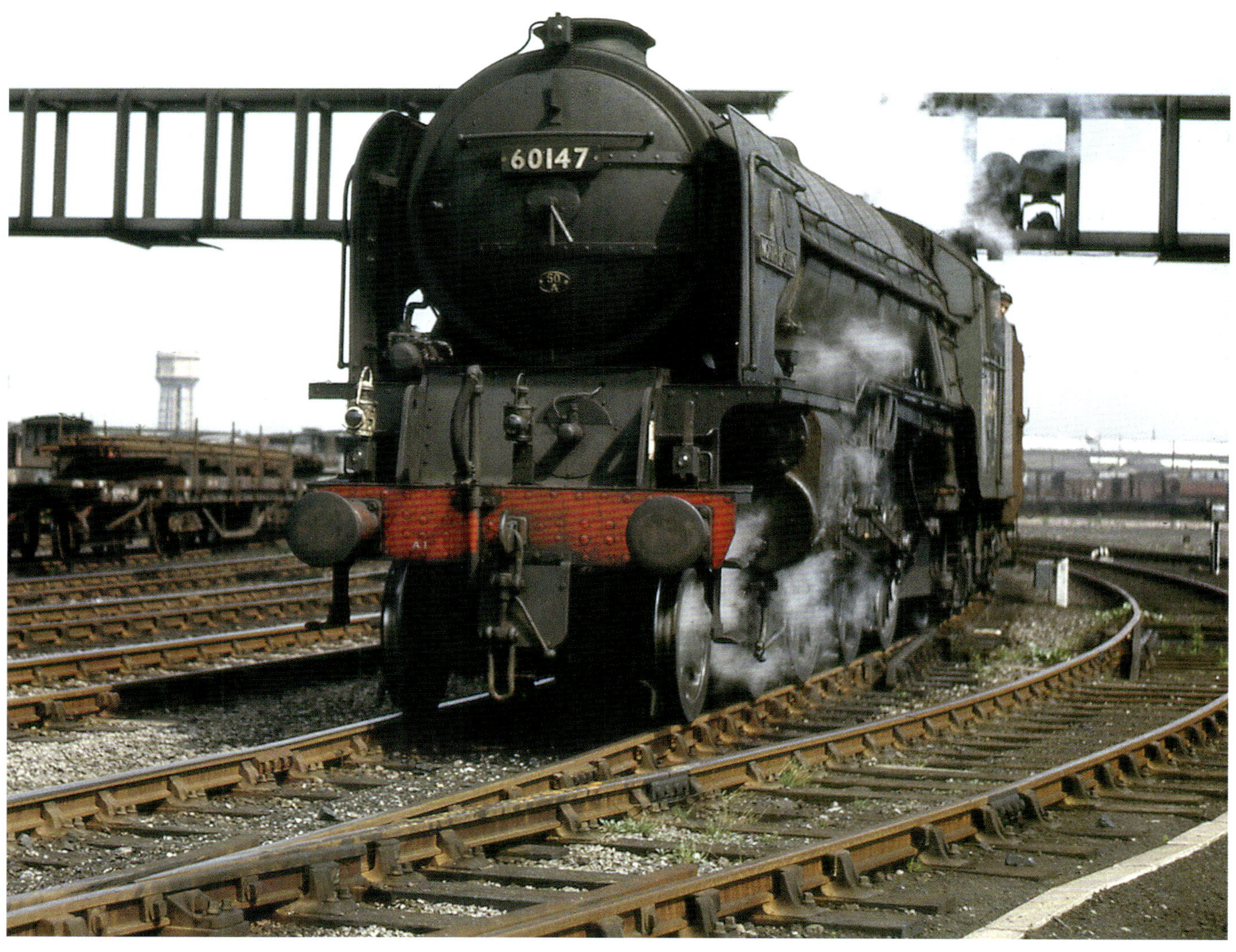

Opposite top: Monday 8 June 1964. A few miles south of York near Dringhouses, Ministry of Supply Austerity Class 8F 2-8-0 No. 90698 is working an 'up' mixed-goods train. Constructed by the Vulcan Foundry in 1944, she would be numbered 79221 by the War Department and enter British Railways stock during December 1948. She would be withdrawn from service during 1967. (Andrew G. Forsyth)

Opposite bottom: Tuesday 9 June 1964. Showing signs of severe priming on the smoke box and boiler, Class K1 2-6-0 No. 62056 is seen at the head of a train of coal hoppers near York station. Constructed by the NBL during 1949, she would only give sixteen years of service before being withdrawn in 1965. (Andrew G. Forsyth)

Above: Tuesday 9 June 1964. Near York station, the impressive frontage of Class A1 4-6-2 No. 60147 *North Eastern* is seen at the head of a goods working. Entering service from Darlington Works during 1949, she would be withdrawn two months after this photograph was taken, in August 1964. (Andrew G. Forsyth)

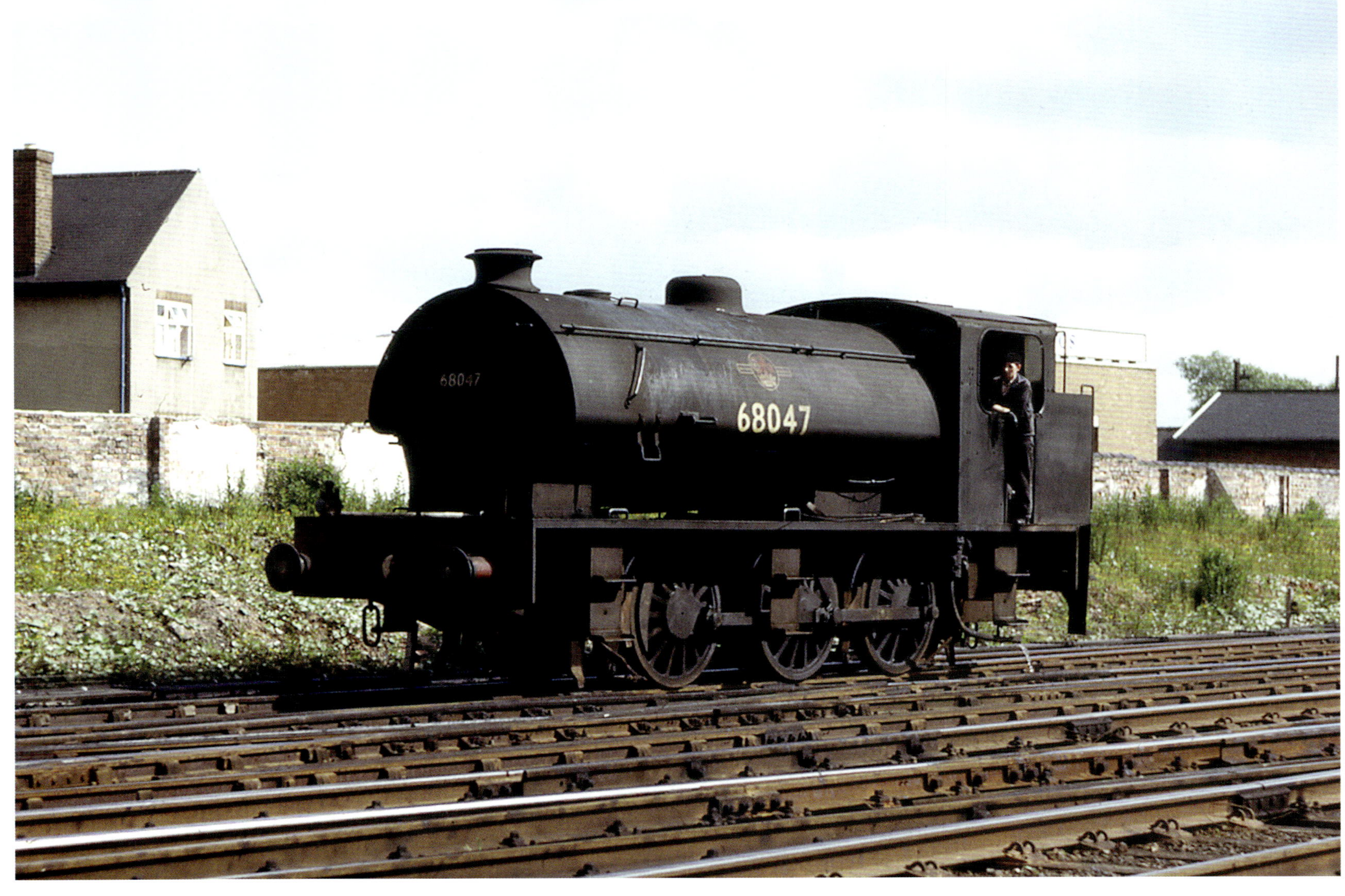

Above: Thursday 11 June 1964. Seen moving light engine near Darlington shed is Class J94 0-6-0 saddle tank No. 68047. Constructed by W.G. Bagnall & Co. during 1945, she was originally numbered 75258 by the War Department and would be purchased by the LNER in 1946. She would be withdrawn from service in 1965. (Andrew G. Forsyth)

Opposite top: Thursday 11 June 1964. Another light engine movement is seen here with Class K1 2-6-0 No. 62048 near Darlington shed. Entering service from the NBL in Glasgow during 1949, she would be withdrawn in 1967. (Andrew G. Forsyth)

Opposite bottom: Thursday 11 June 1964. At Darlington shed ex-NER Class P3 (LNER Class J27) 0-6-0 No. 65880 is seen in ex-works condition. A product of the same works in 1922, she would be withdrawn from service during 1967. (Andrew G. Forsyth)

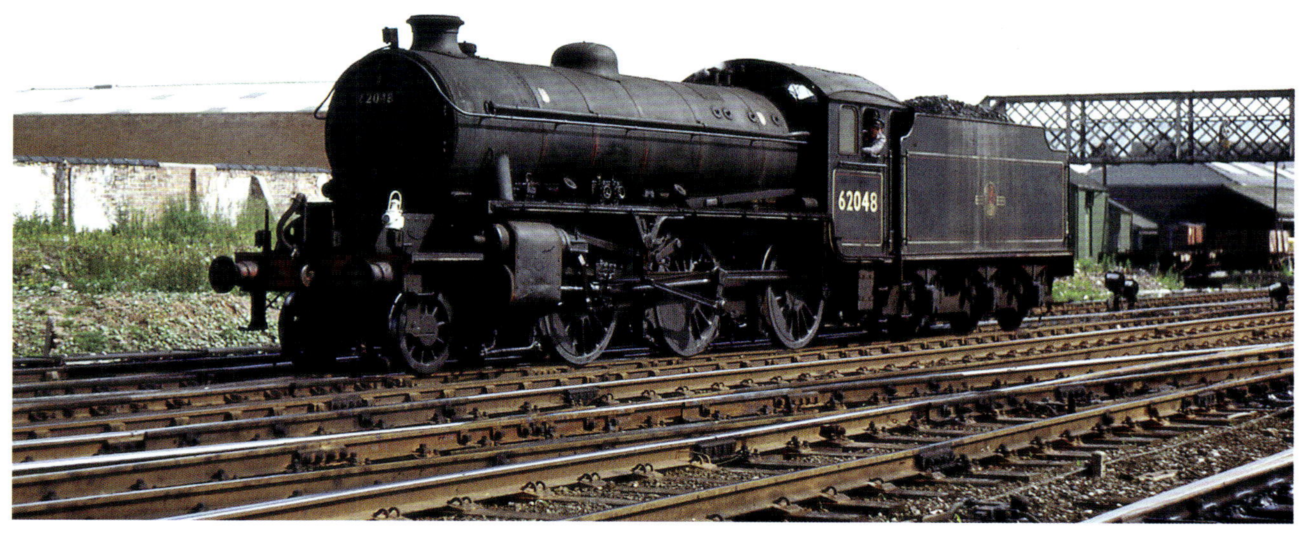

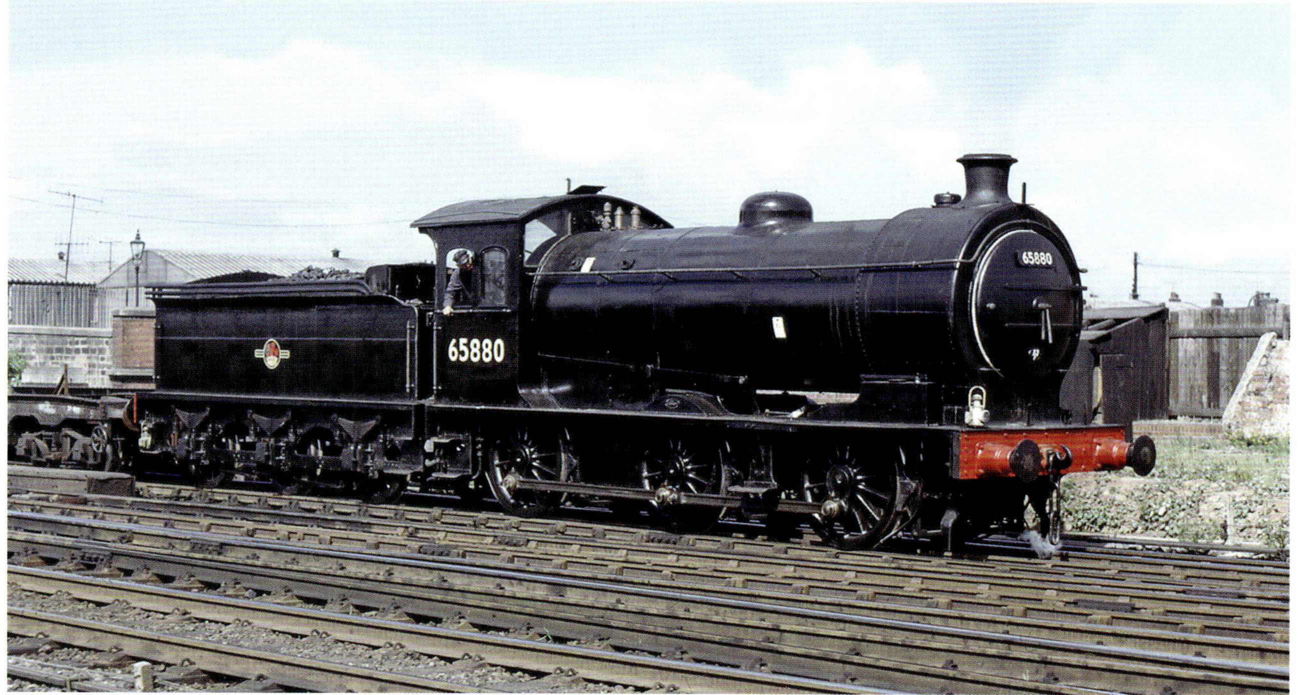

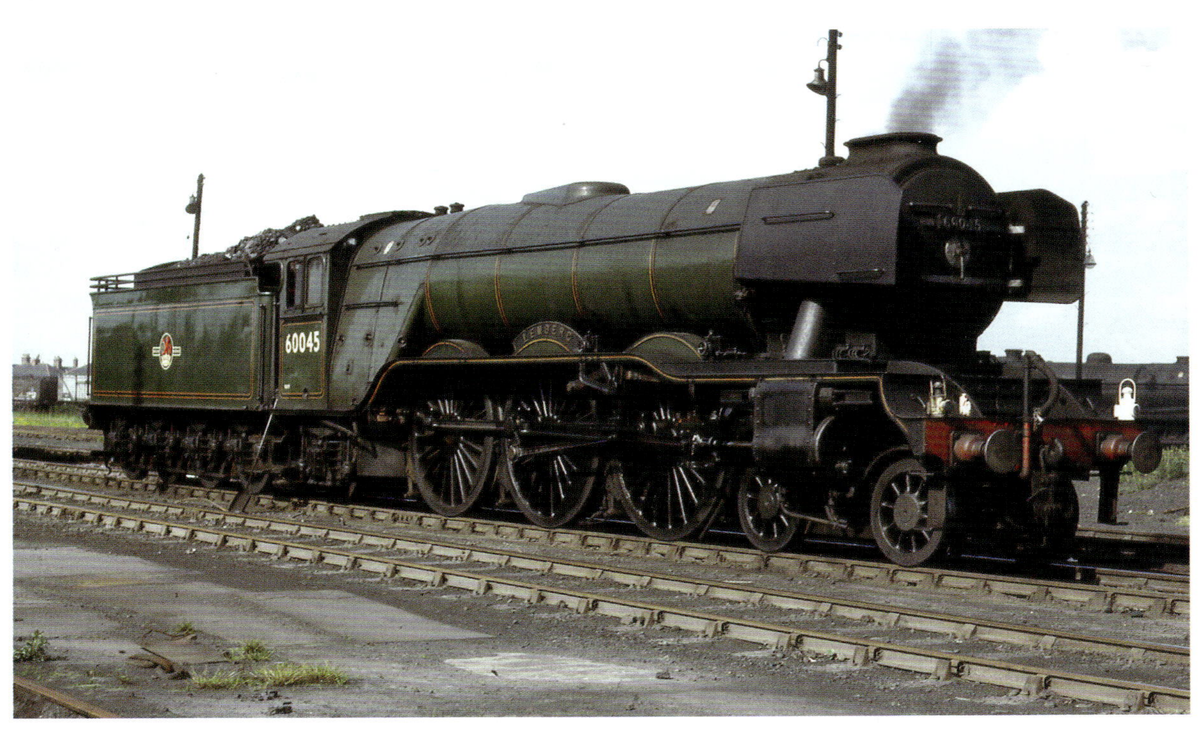

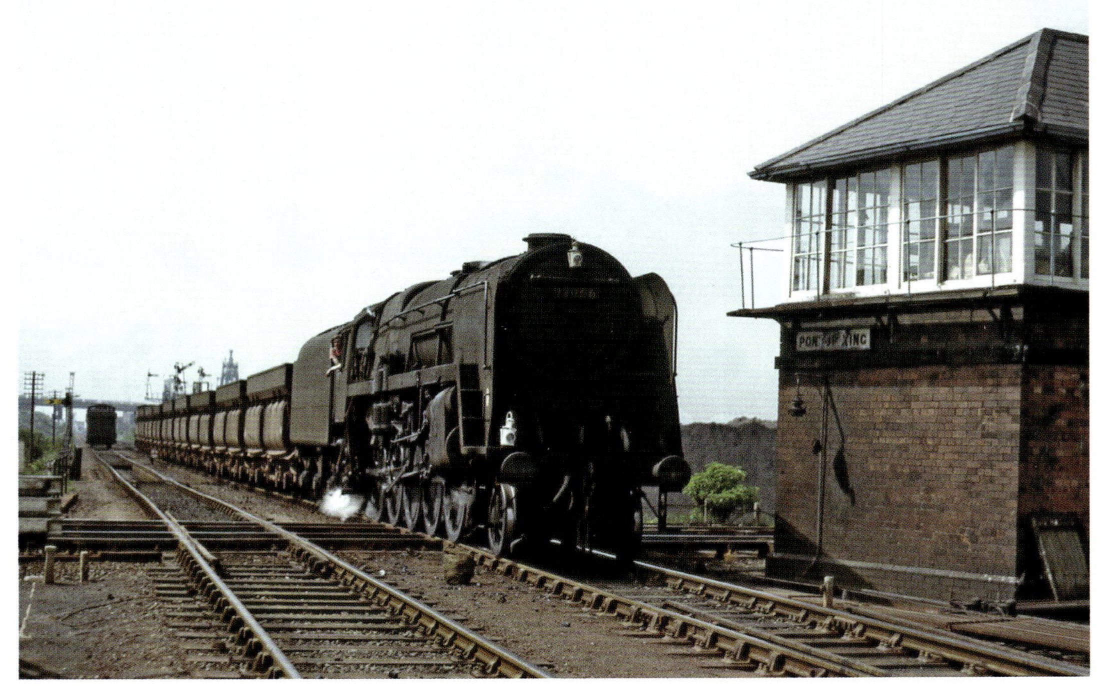

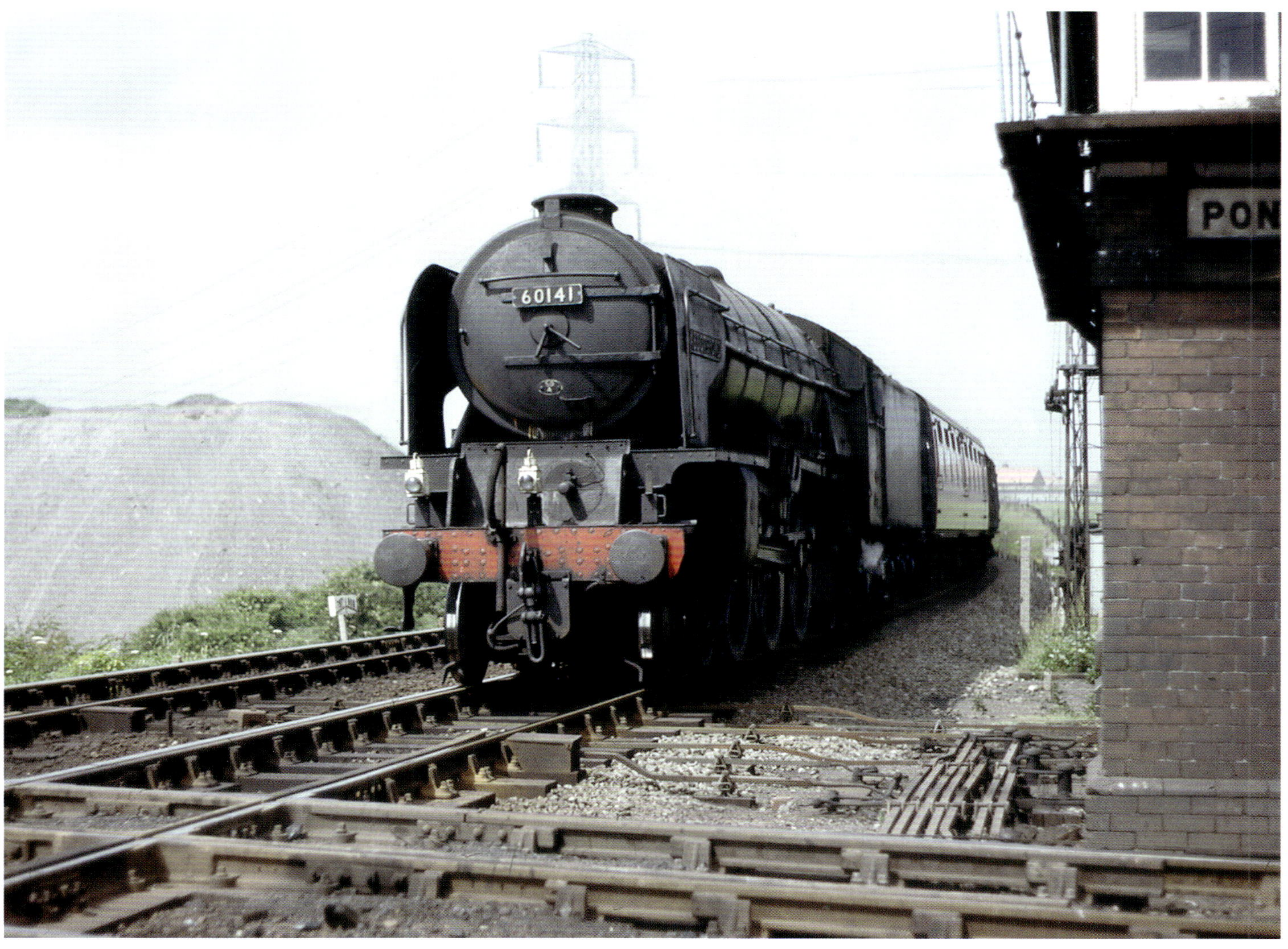

Opposite top: Thursday 11 June 1964. Also at Darlington shed is what the photographer has noted as the stand-by locomotive for the day. Ex-LNER Class A3 4-6-2 No. 60045 *Lemberg* is looking in good, clean condition. Constructed at Doncaster Works during 1924 as a Class A1 locomotive, she would be rebuilt in 1927 as a Class A3. Named after the racehorse that won the 1910 Derby, she would be fitted with a double chimney in 1959 and trough-style smoke deflectors in 1962, before being withdrawn from service five months after this photograph was taken. (Andrew G. Forsyth)

Opposite bottom: Friday 12 June 1964. At Pontop Crossing, BR Standard Class 9F 2-10-0 No. 92066 is seen at the head of a return working to Tyne Dock of empty iron-ore hopper wagons. Constructed at Crewe Works in 1955, she was allocated new to Tyne Dock and would be withdrawn from service there ten years later, during 1965. (Andrew G. Forsyth)

Above: Friday 12 June 1964. At Pontop Crossing, Class A1 4-6-2 No. 60141 *Abbotsford* is hauling a rake of empty coaching stock. Entering service from Darlington Works during 1948 and named after the home of Sir Walter Scott, which is near Galashiels, she would be withdrawn four months after this photograph was taken. (Andrew G. Forsyth)

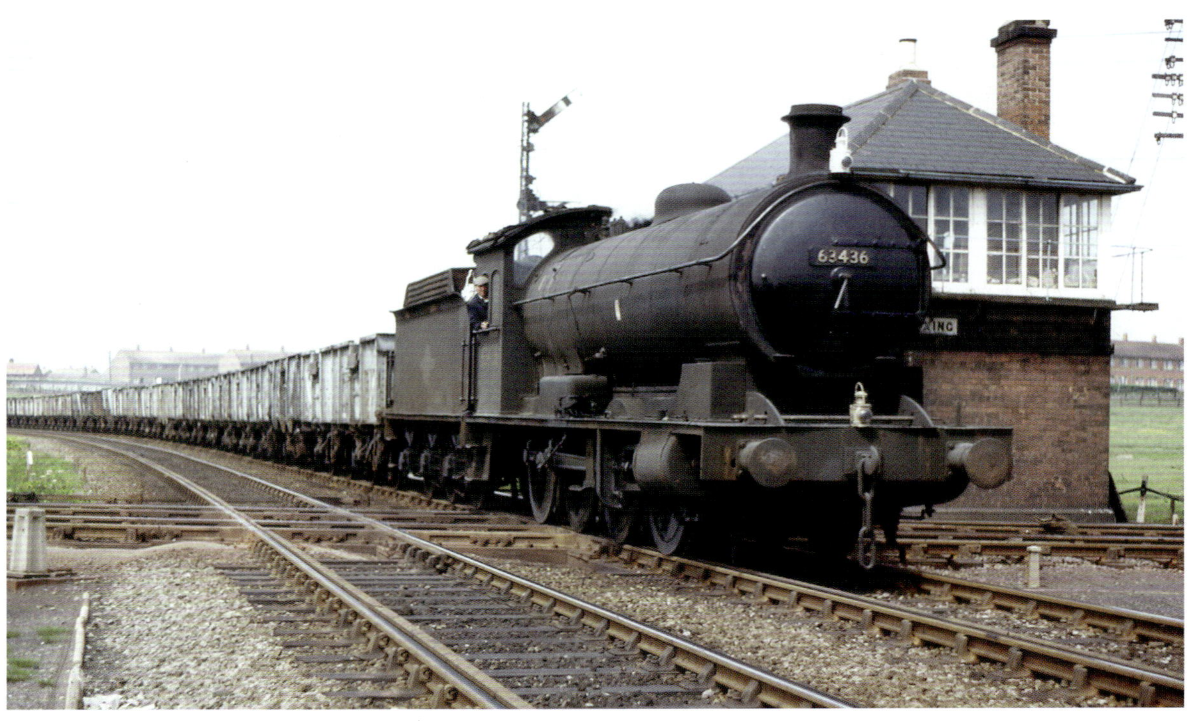

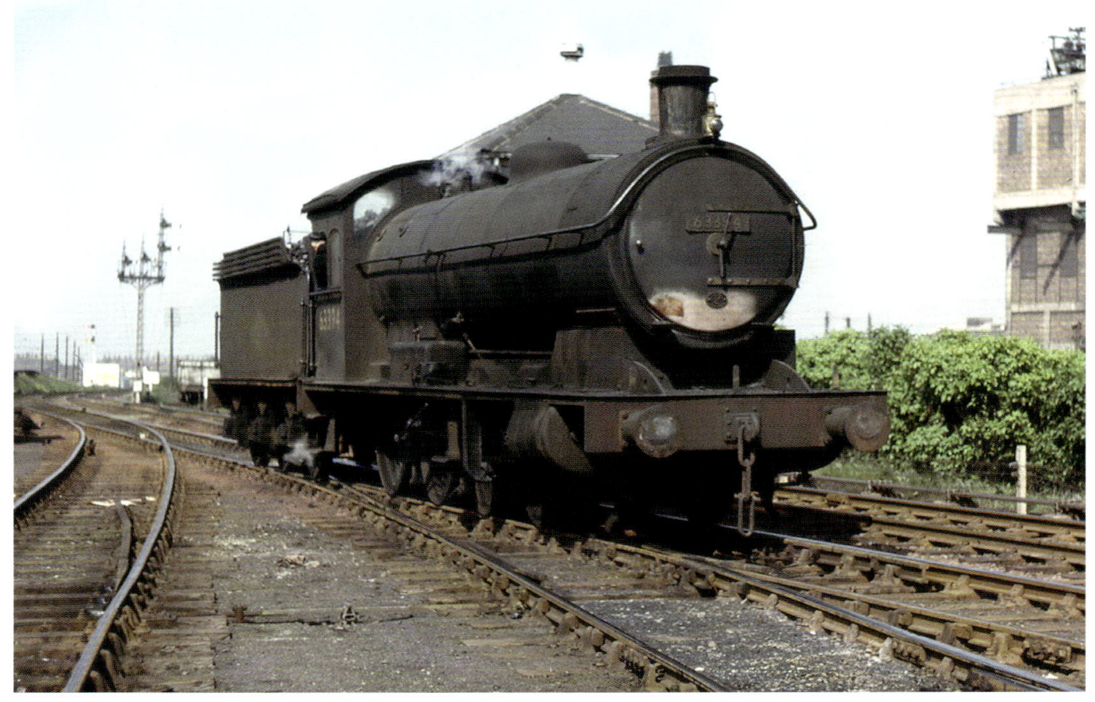

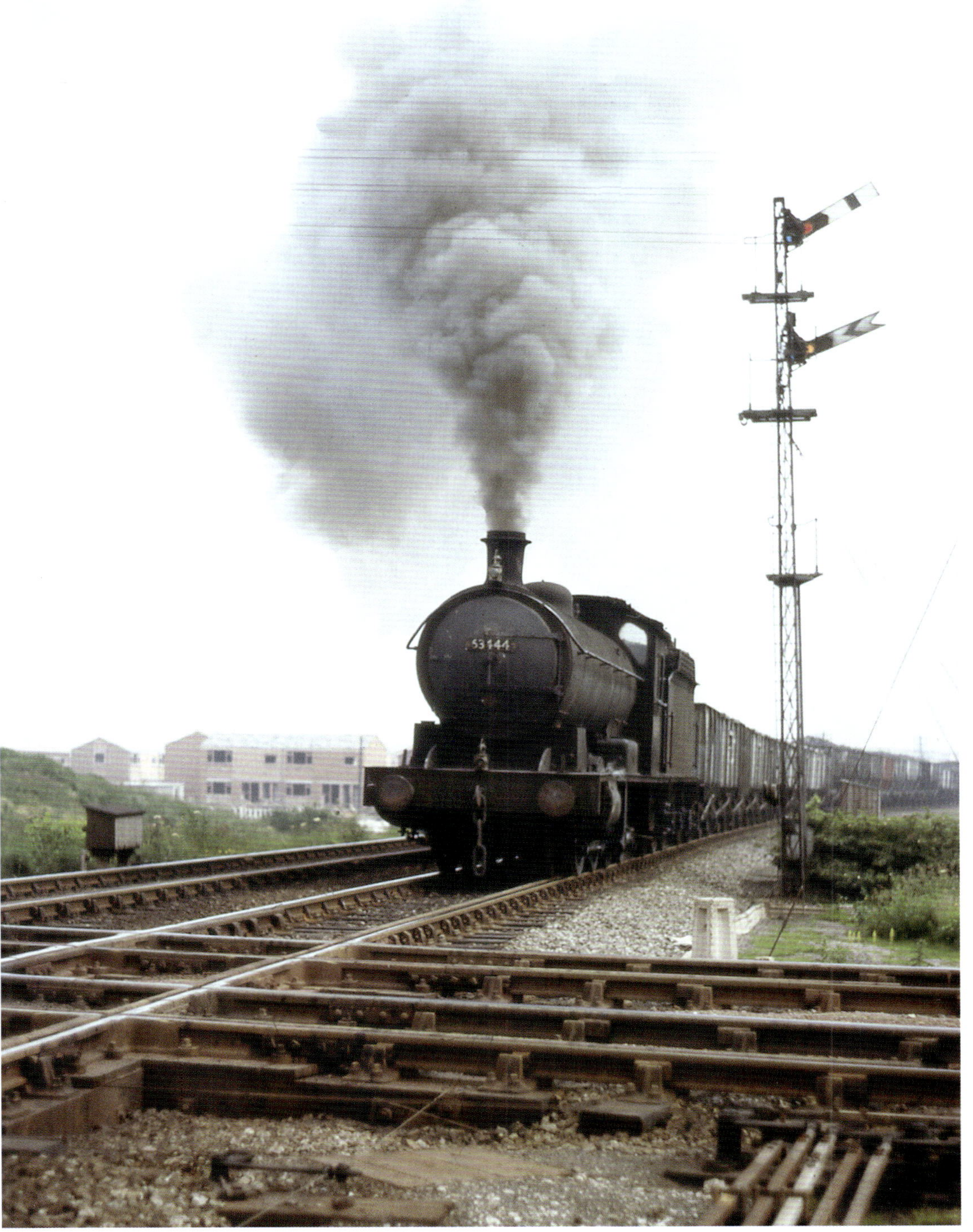

Opposite top: Friday 12 June 1964. Also seen at Pontop Crossing is ex-NER Class T2 (LNER Class Q6) 0-8-0 No. 63436 at the head of a train of empty mineral wagons heading in the direction of Sunderland. Constructed by Armstrong Whitworth & Co. in 1920, she would be withdrawn during 1967. (Andrew G. Forsyth)

Opposite bottom: Friday 12 June 1964. Near Consett, ex-NER Class T2 (LNER Class Q6) 0-8-0 No. 63394 is seen proceeding light engine. Constructed at Darlington Works in 1918, she would be withdrawn during 1967. (Andrew G. Forsyth)

Right: Friday 12 June 1964. This impressive image shows ex-NER Class T2 (LNER Class Q6) 0-8-0 No. 63444 working hard at the head of a train of loaded mineral wagons at Pontop Crossing, heading toward Pelaw. Entering service from Armstrong Whitworth & Co. in 1920, she would be withdrawn during 1965. (Andrew G. Forsyth)

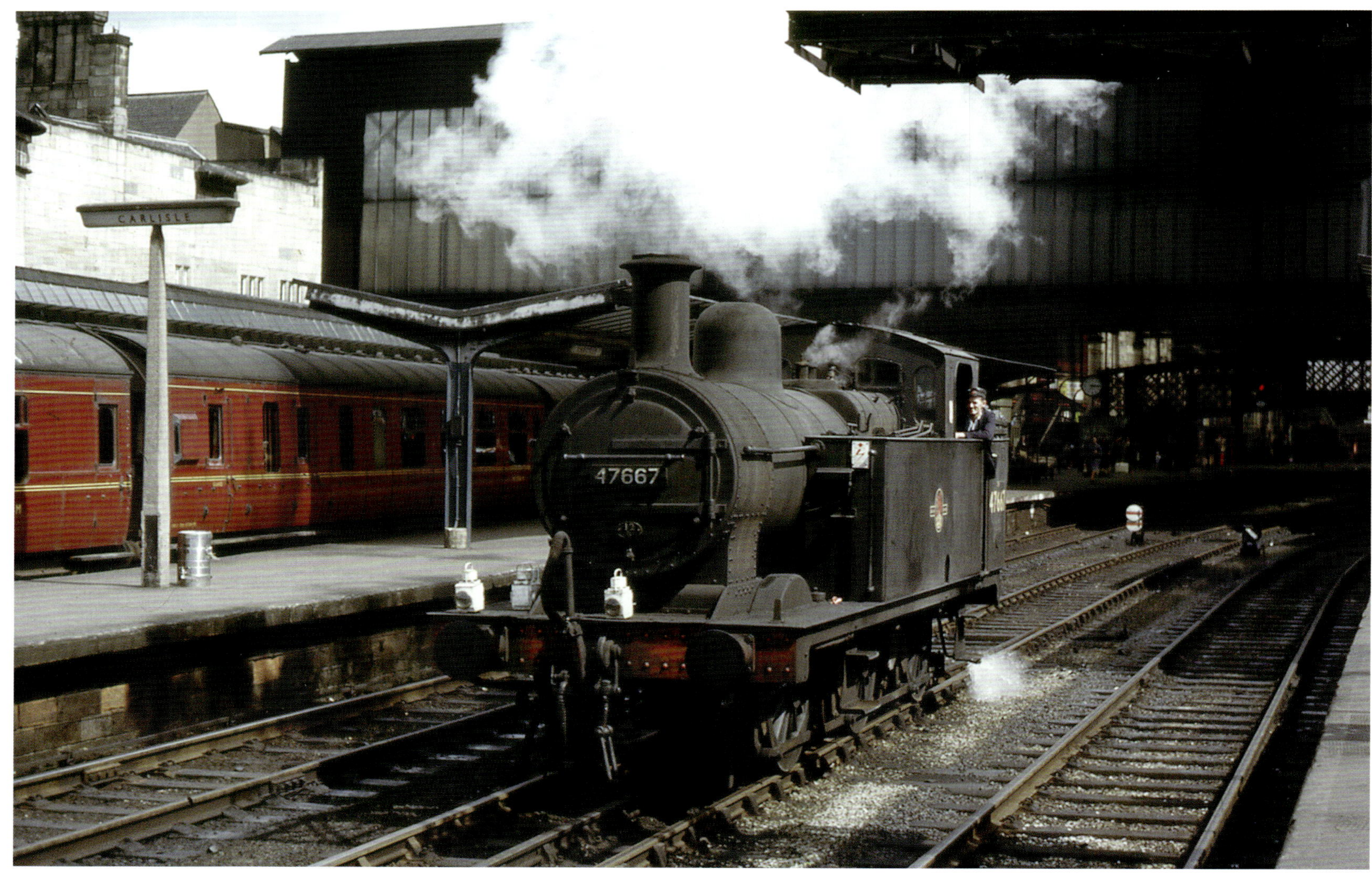

Above: Saturday 13 June 1964. At Carlisle station, the south-end pilot locomotive for the day is ex-LMS Class 3F 0-6-0 tank No. 47667, and she is seen moving about her duties. Entering service from Horwich Works during 1931, she would be withdrawn in 1966. (Andrew G. Forsyth)

Opposite top: Monday 15 June 1964. Seen passing Craigentinny signal box, to the east of Edinburgh, at the head of a five-coach 'up' working is ex-LMS Class 5 4-6-0 No. 44975 bearing a 64C Dalry Road shed code. Constructed at Crewe Works during 1946 and allocated new to Perth shed, she would end her days based at Dalry Road and be withdrawn from service in 1965. (Andrew G. Forsyth)

Opposite bottom: Wednesday 17 June 1964. With a long train of 16-ton mineral wagons behind the tender, ex-NBR Class S (LNER Class J37) 0-6-0 No. 64595 is seen passing the Rothes Colliery at Thornton. A product of the NBL during 1919, she would be withdrawn in 1966. (Andrew G. Forsyth)

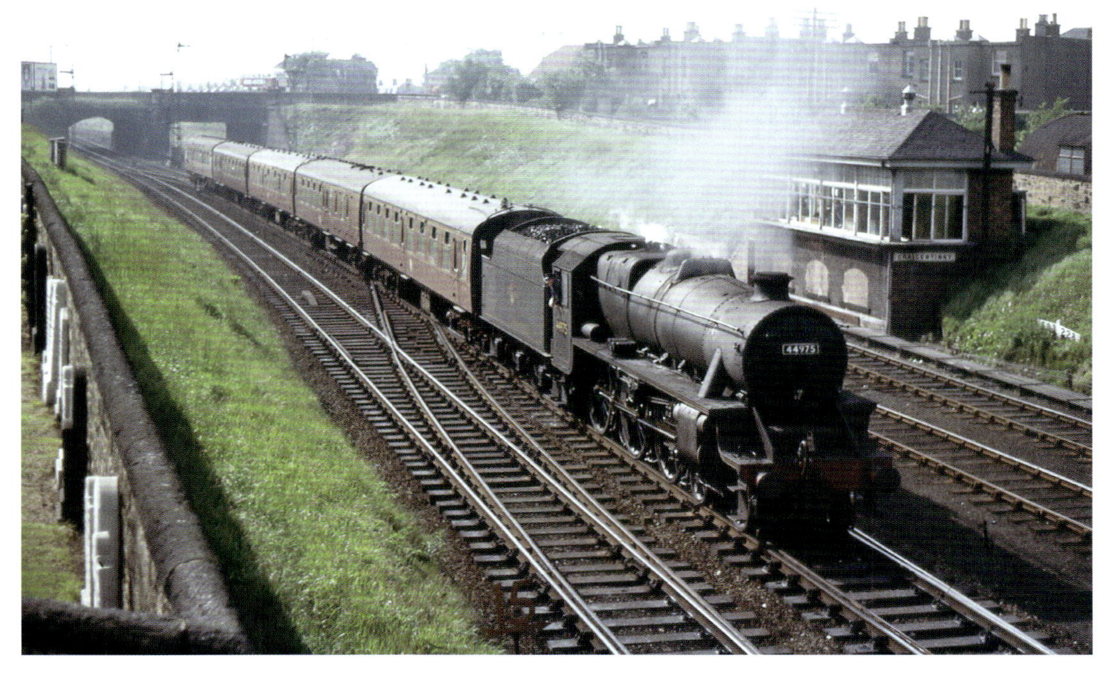

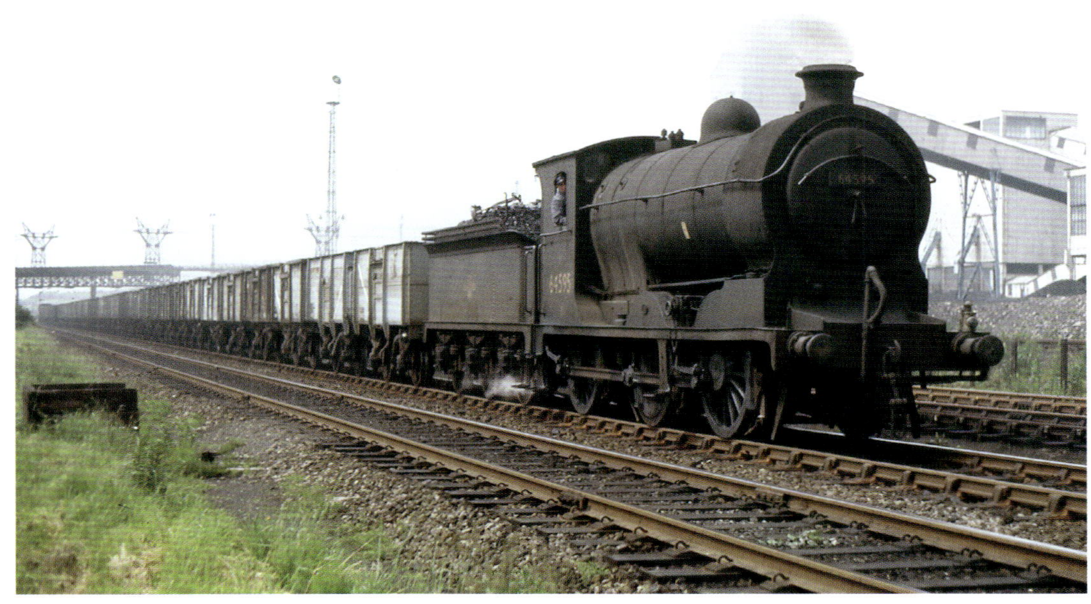

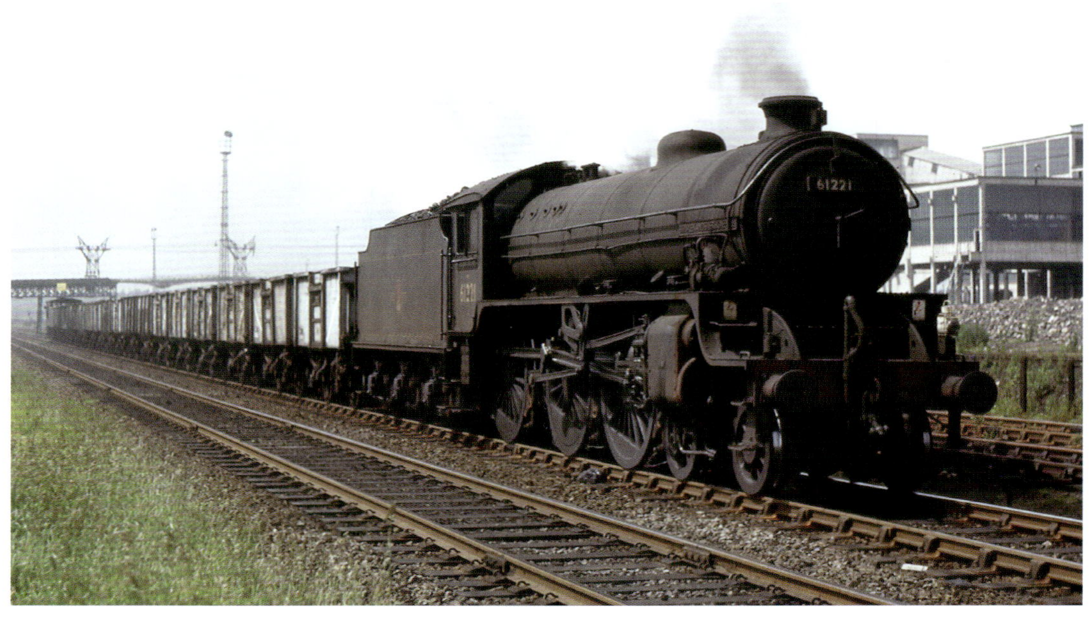

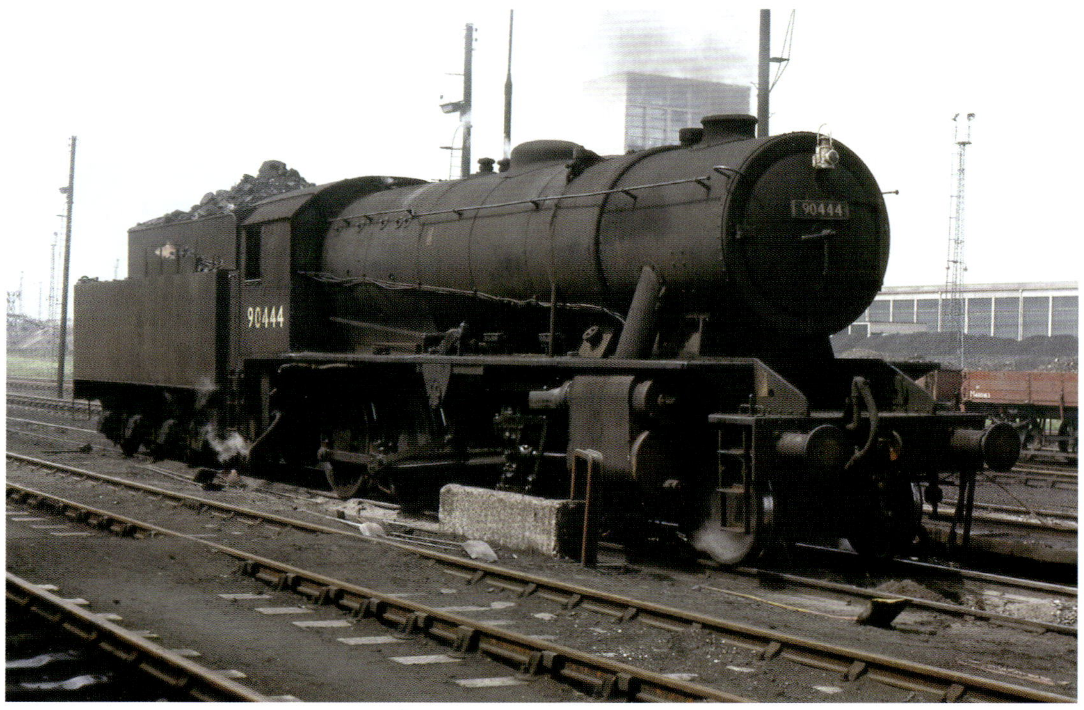

Opposite top: Wednesday 17 June 1964. Another long train of 16-ton mineral wagons is seen here passing the Rothes Colliery at Thornton behind ex-LNER Class B1 4-6-0 No. 61221 *Sir Alexander Erskine-Hill*. Constructed by the NBL in 1947 and named after the Member of Parliament for Edinburgh North, she would be withdrawn during 1965. (Andrew G. Forsyth)

Opposite bottom: Wednesday 17 June 1964. At Thornton Junction shed Ministry of Supply Austerity Class 8F 2-8-0 No. 90444 is fully coaled up and awaiting her next duty. Constructed by the Vulcan Foundry during 1944 and numbered 77505 by the War Department, she would be loaned to the LNER in 1945 and purchased by them in 1946 and numbered 3123. Initially becoming No. 63123 with British Railways, she would be allocated to Dundee shed but end her days at Thornton Junction, being withdrawn during 1967. (Andrew G. Forsyth)

Right: Friday 19 June 1964. At Bathgate shed, ex-NBR Class S (LNER Class J37) 0-6-0 No. 64614 is parked just inside the shed entrance. A product of the NBL in 1920, she would be withdrawn six months after this photograph was taken, in December 1964. (Andrew G. Forsyth)

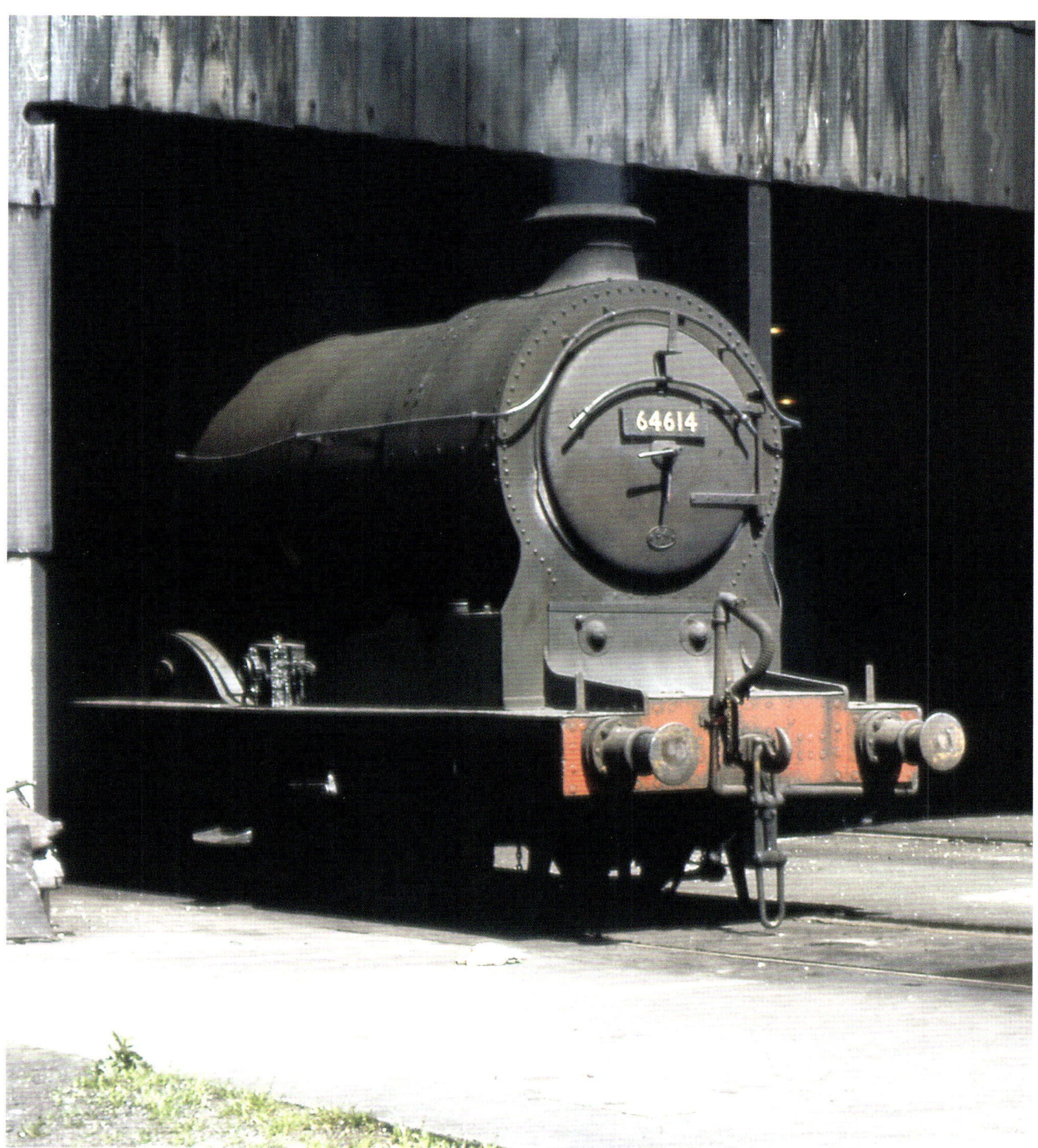

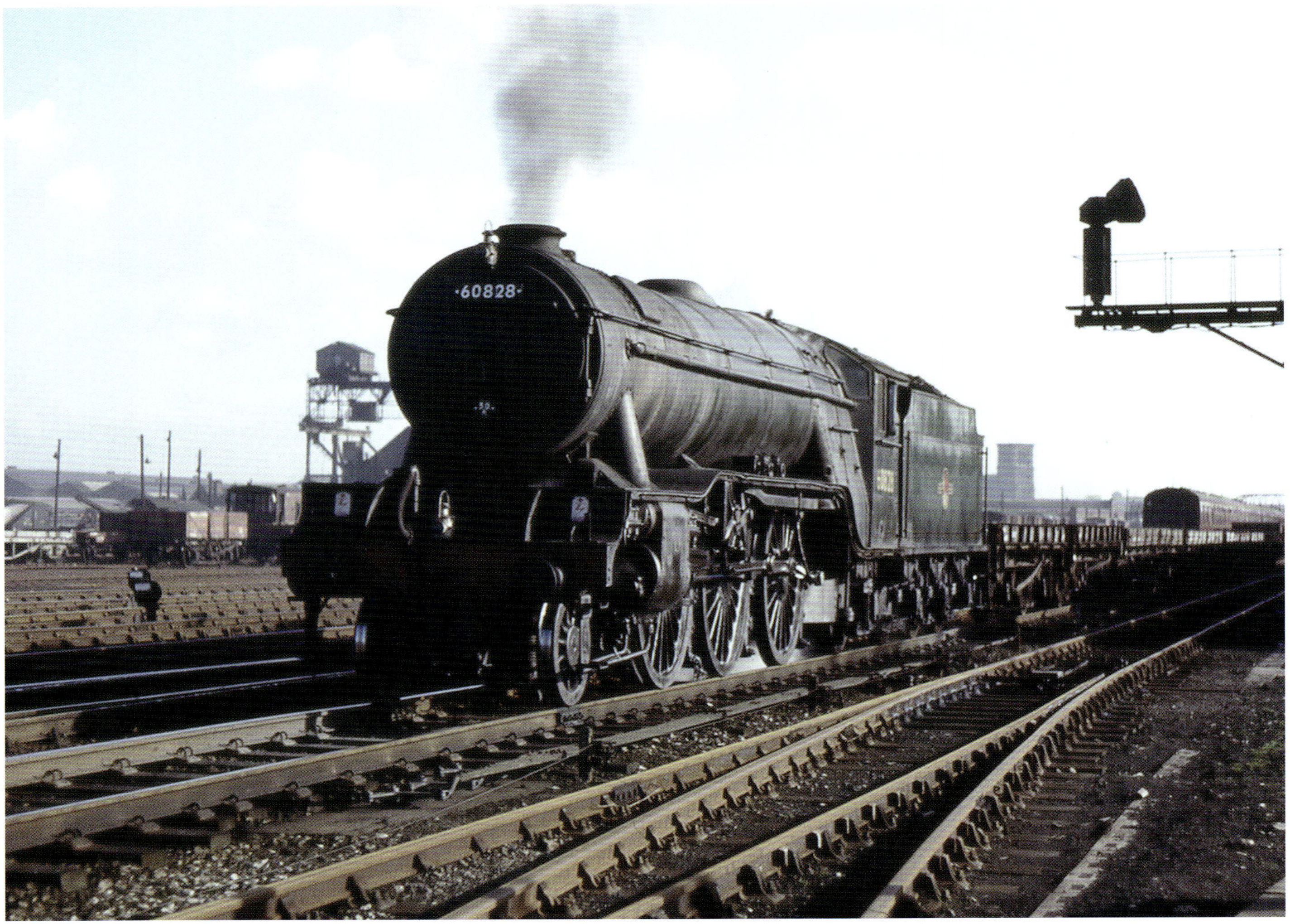

Above: Sunday 11 April 1965. With a 'down' goods train working seen approaching Darlington, ex-LNER Class V2 2-6-2 No. 60828 is in good, clean condition. Constructed at Darlington Works during 1938, she would be rebuilt in 1958 with separate cylinder blocks, as seen here, with the use of outside steam pipes. Allocated new to York shed, she would spend some time during the 1950s allocated to Kings Cross but by 1958 was back at York, whose 50A shed code she is carrying. She would be withdrawn from service six months after this photograph was taken, in October 1965. (Andrew G. Forsyth)

Opposite top: Sunday 25 April 1965. The photographer's notes show that this Tyne Dock–Consett iron-ore train is the 2.50 p.m. working, which is seen near Bank Top with BR Standard Class 9F 2-10-0 No. 92097 having steam to spare on the climb. Entering service from Crewe Works in 1956 and allocated new to Tyne Dock shed, she spent her entire ten-year working life based there, being withdrawn during October 1966. (Andrew G. Forsyth)

Opposite bottom: Sunday 25 April 1965. In the same position as in the previous photograph, ex-NER Class T2 (LNER Class Q6) 0-8-0 No. 63398 is seen at the head of a mixed train of mineral wagons. A product of Darlington Works during 1918, she would be withdrawn from service six months after this photograph was taken, in October 1965. (Andrew G. Forsyth)

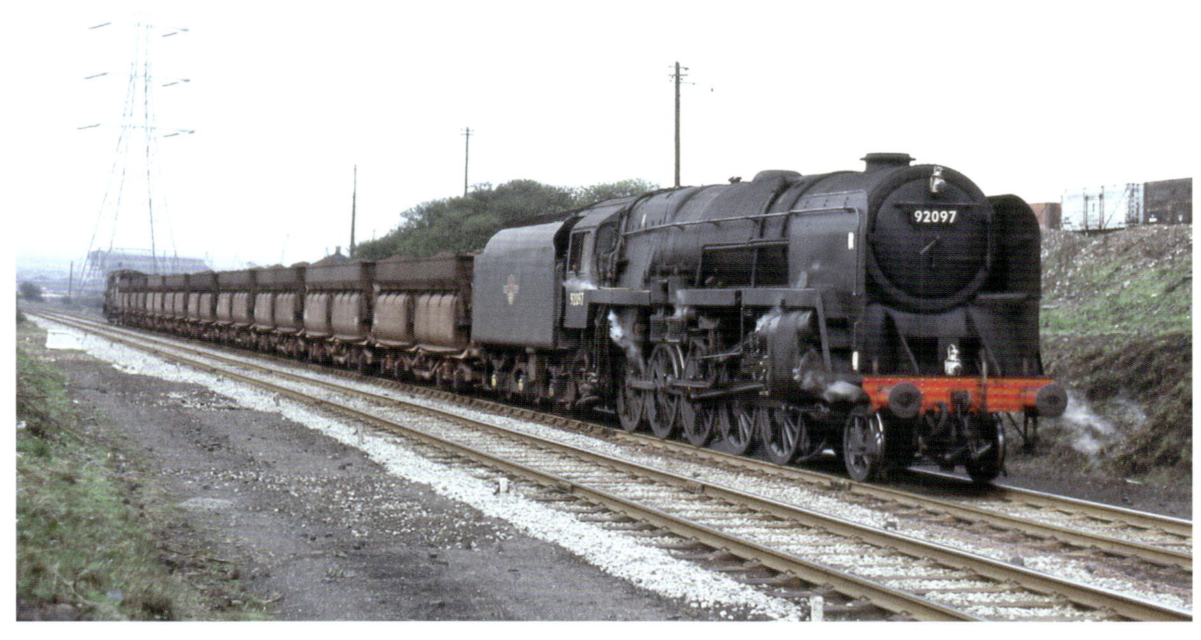

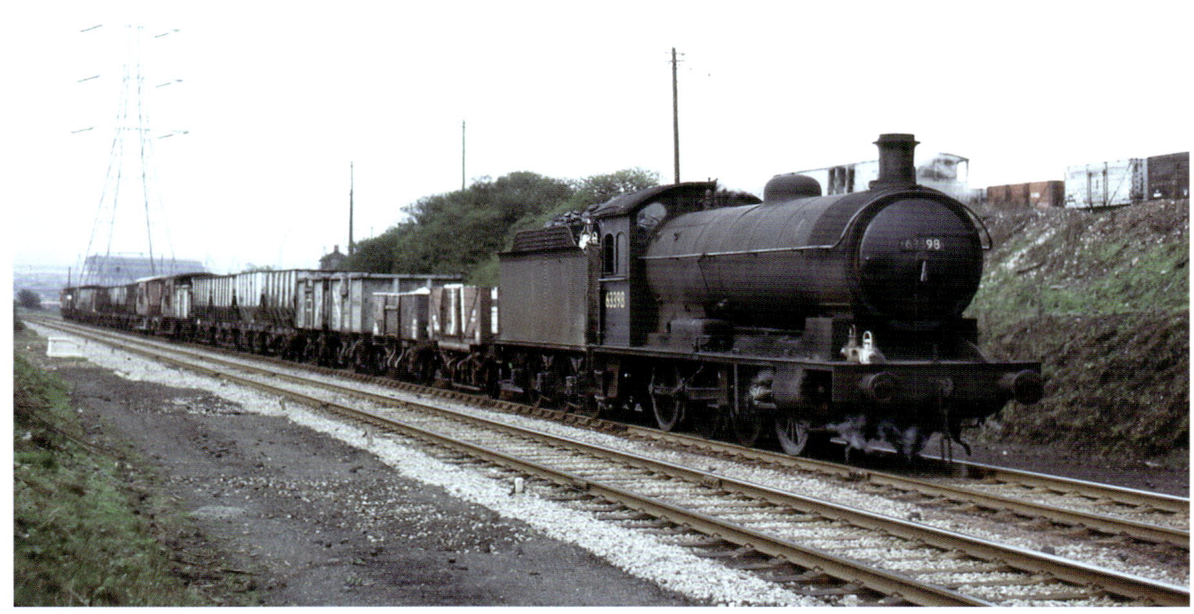

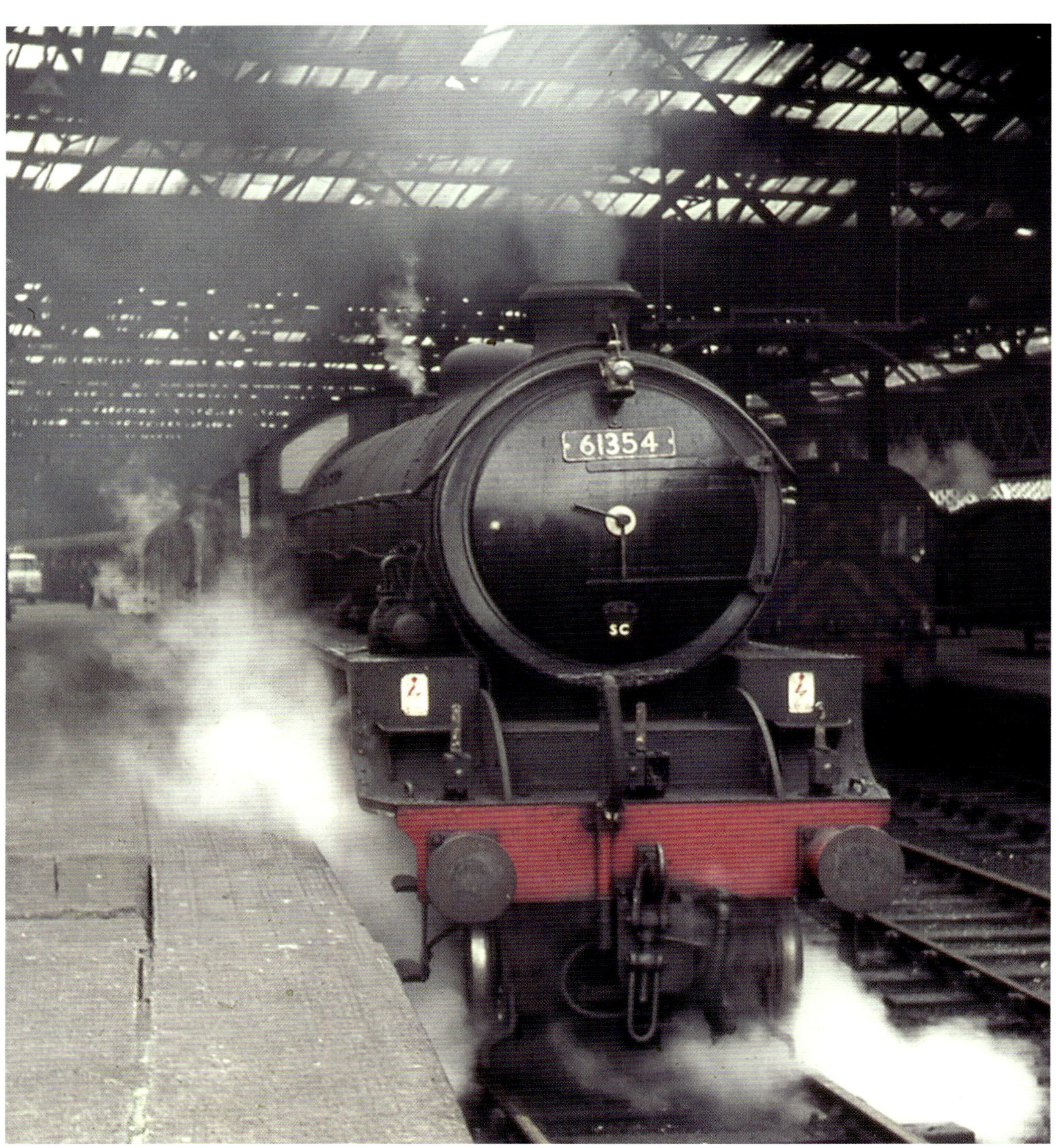

Wednesday 12 May 1965. At Waverley station in Edinburgh, Class B1 4-6-0 No. 61354 is preparing to depart with the 4.10 p.m. 'all stations' passenger working to Hawick, which was timetabled to arrive there at 6.00 p.m. Constructed at Darlington Works during 1949, she would be allocated new to St Margarets in Edinburgh but end her days based at Dundee Tay Bridge shed, being withdrawn in 1967. (Anthony Scarsbrook)

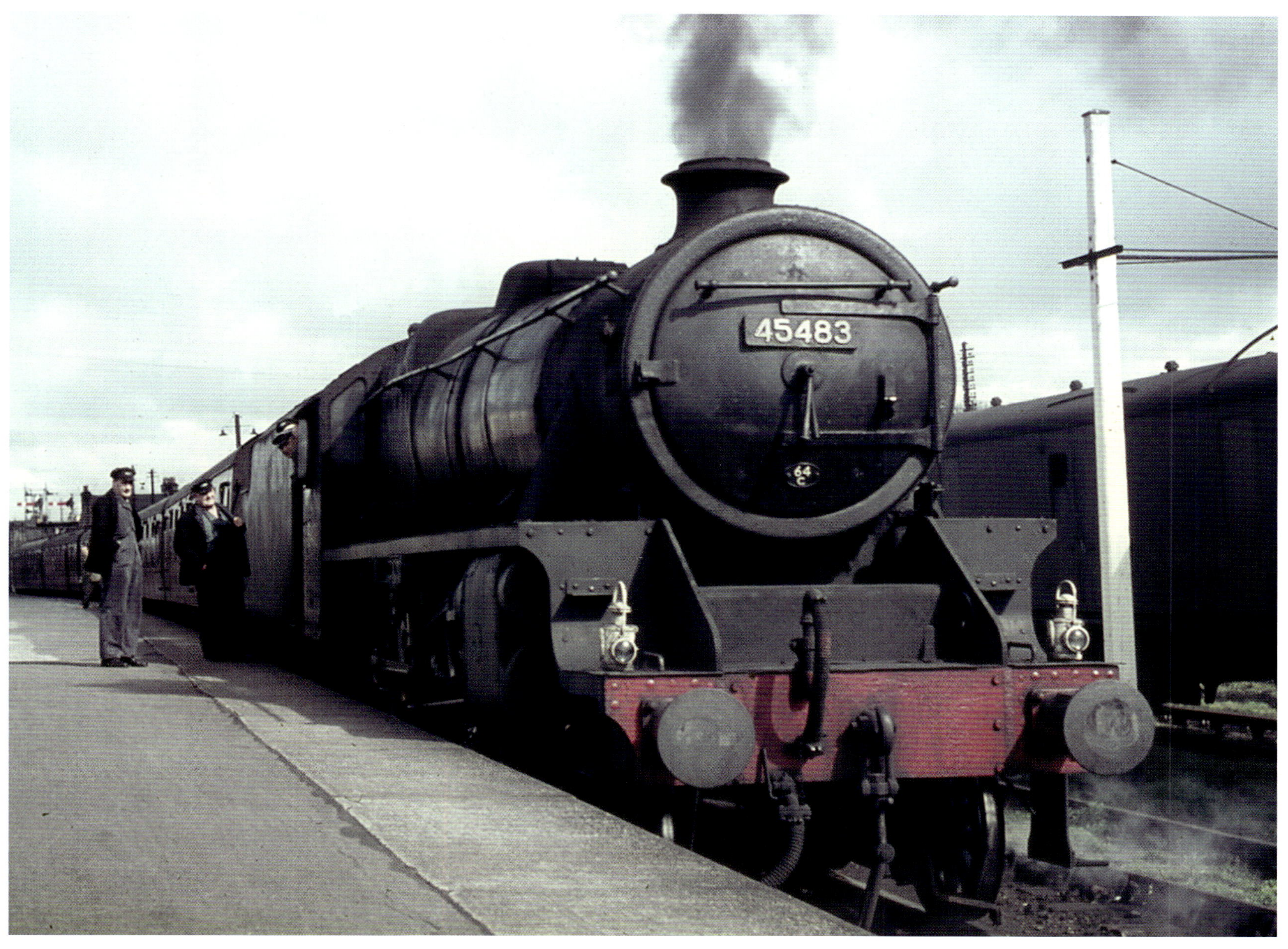

Saturday 15 May 1965. Waiting to depart from Stirling station in charge of the 10.15 a.m. working to Edinburgh Waverley via Larbert is ex-LMS Class 5 4-6-0 No. 45483, bearing a 64C Dalry Road shed code. Constructed at Derby Works during 1943, she would spend her entire working life based at Scottish sheds and be withdrawn in 1966 while based at St Margarets in Edinburgh. (Anthony Scarsbrook)

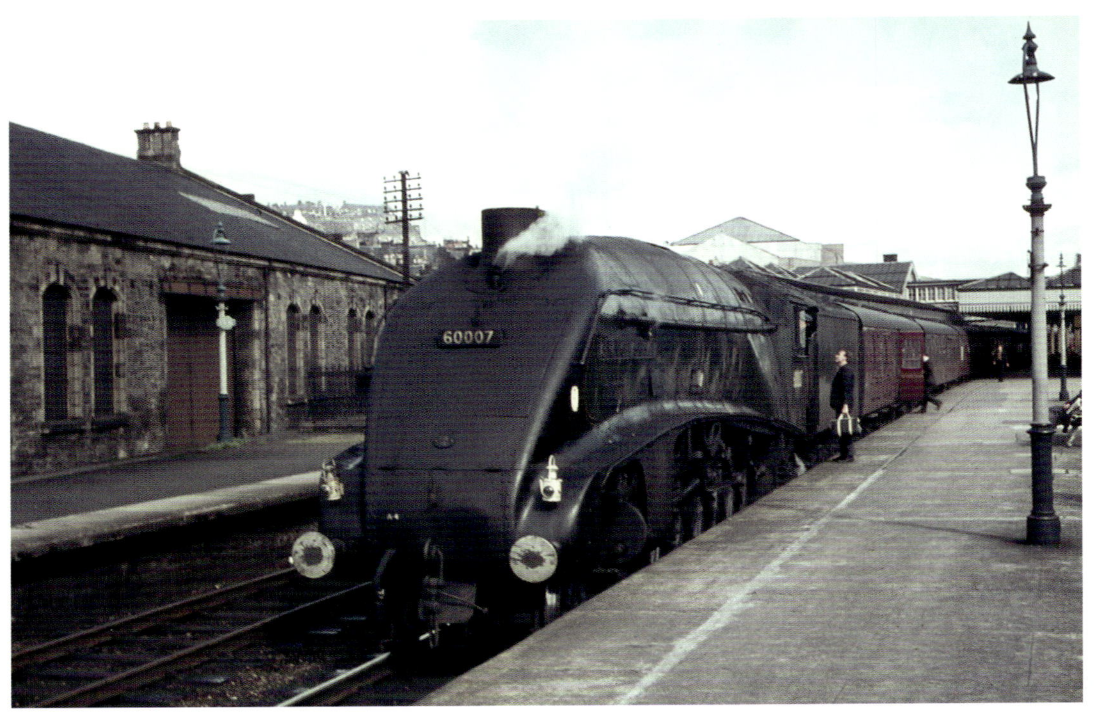

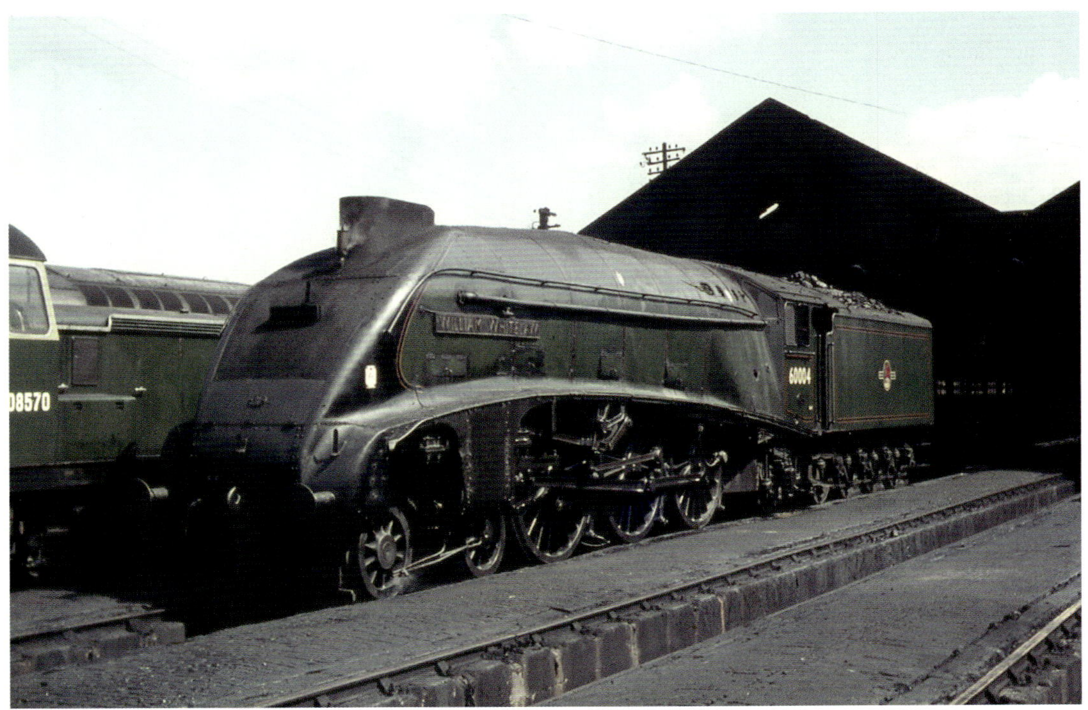

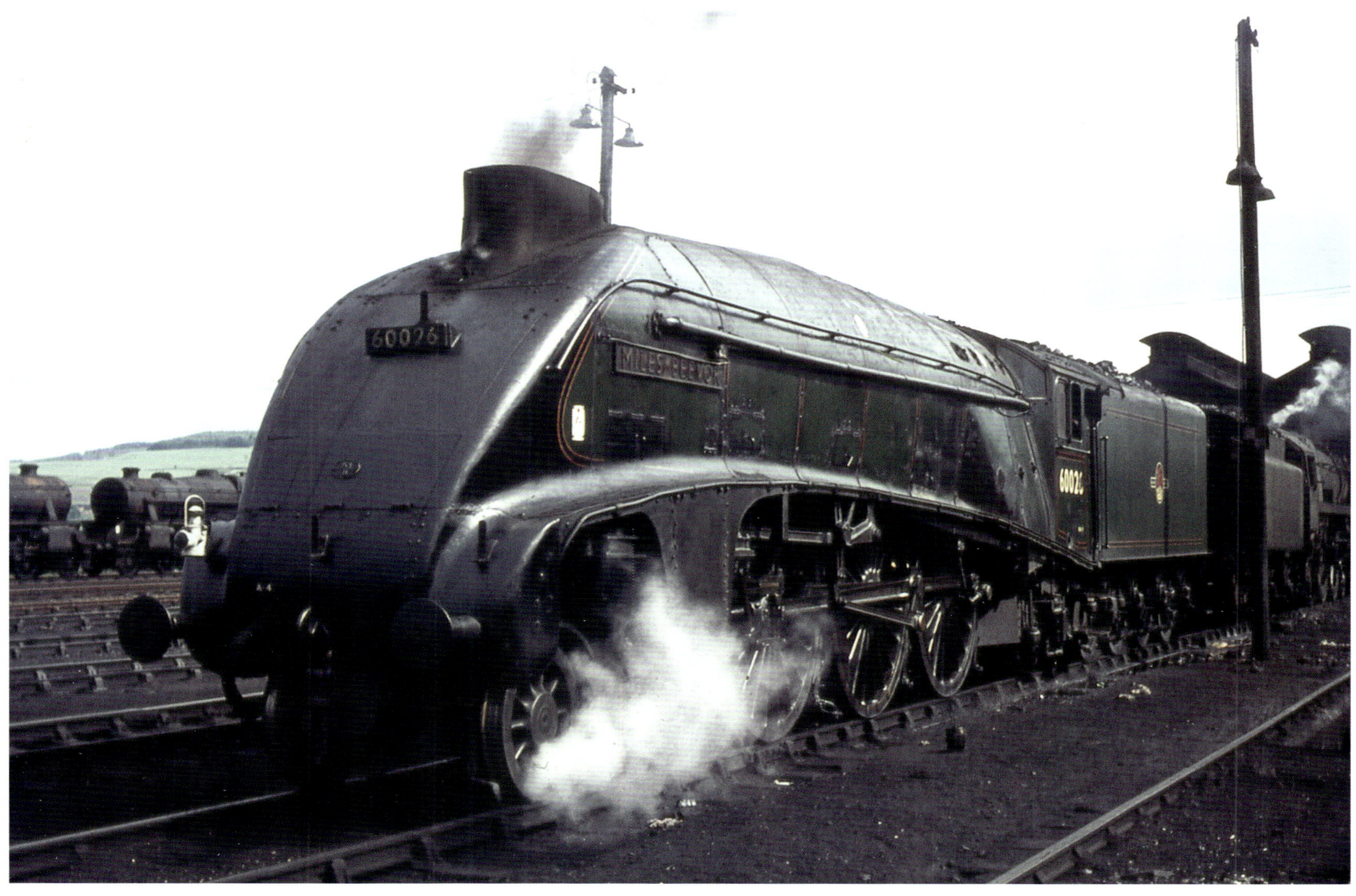

Opposite top: Saturday 15 May 1965. At Stirling station ex-LNER Class A4 4-6-2 No. 60007 *Sir Nigel Gresley* is waiting to depart with a Glasgow-bound passenger train. Constructed at Doncaster Works during 1937 and named after her designer, she would end her days based at Ferryhill shed in Aberdeen, being withdrawn in 1966. Purchased by a preservation society and seen working many main-line specials, she is currently undergoing a heavy overhaul. (Anthony Scarsbrook)

Above: Sunday 16 May 1965. Another Class A4 is seen here, on this occasion at Perth shed. No. 60026 *Miles Beevor* had entered service from Doncaster Works during 1937 and was originally named *Kestrel*, the name being changed in 1947 to honour the chief legal adviser to the LNER. Also in good, clean condition, she would end her working life based at Ferryhill shed and be withdrawn at the end of 1965. (Anthony Scarsbrook)

Opposite bottom: Saturday 15 May 1965. At St Margarets shed in Edinburgh ex-LNER Class A4 4-6-2 No. 60004 *William Whitelaw* is looking relatively clean. (Anthony Scarsbrook)

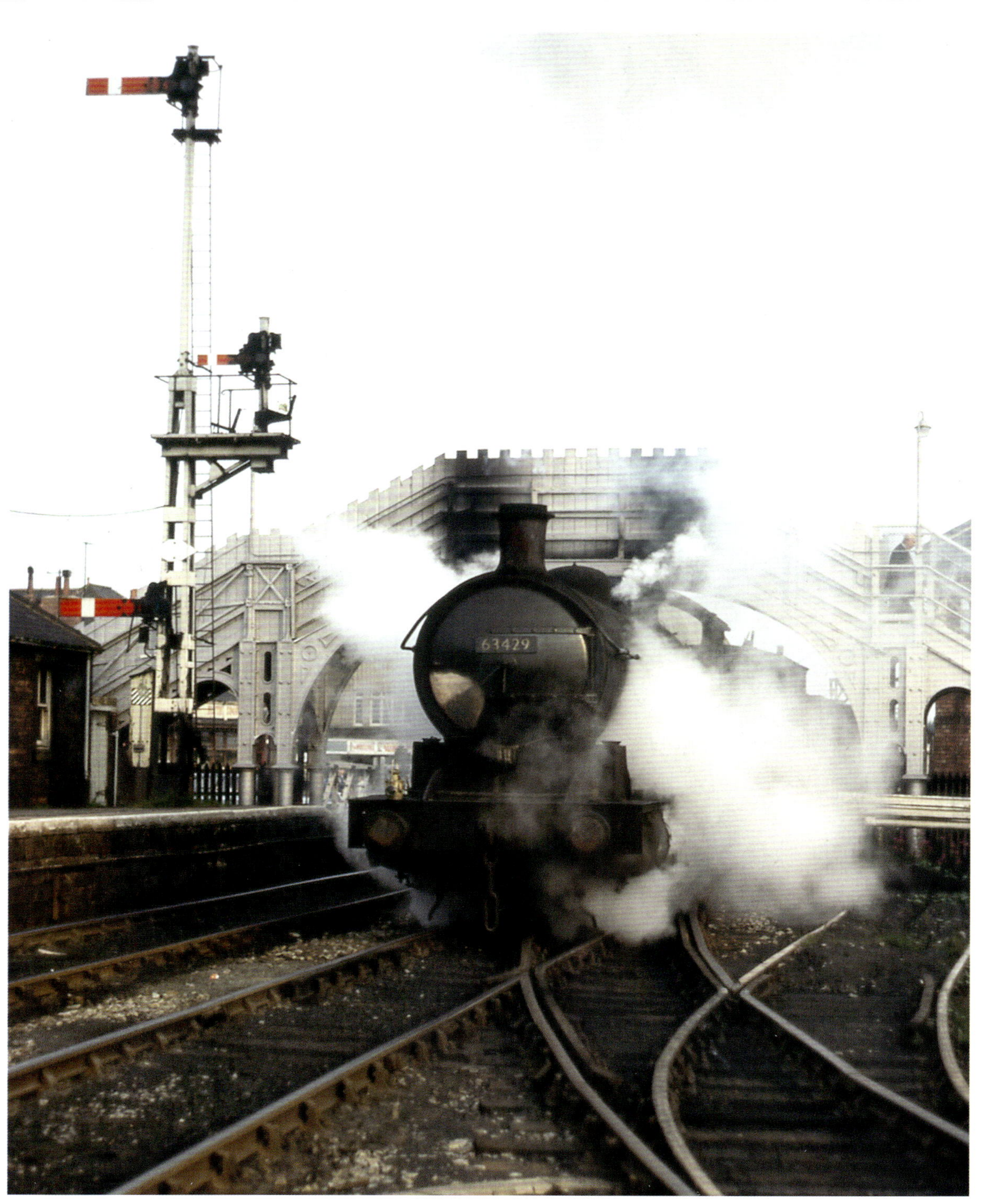

Left: Wednesday 11 August 1965. Ex-NER Class T2 (LNER Class Q6) 0-8-0 No. 63429 is seen passing Bedlington station with what can only be described as seriously leaking glands. A product of Armstrong Whitworth & Co. in 1920, she would be withdrawn during 1967. (Andrew G. Forsyth)

Opposite top: Wednesday 11 August 1965. Seen arriving at Bedlington sidings is ex-NER Class P3 (LNER Class J27) 0-6-0 No. 65804 in charge of a train of open wagons. Constructed at Darlington Works during 1908, she would be withdrawn in 1967. (Andrew G. Forsyth)

Opposite bottom: Friday 13 August 1965. In the typical work-stained condition of the period, ex-NER Class T2 (LNER Class Q6) 0-8-0 No. 63445 is at the head of a train of coal hopper wagons near Bensham. A product of Armstrong Whitworth & Co. in 1920, she would be withdrawn from service during 1966. (Andrew G. Forsyth)

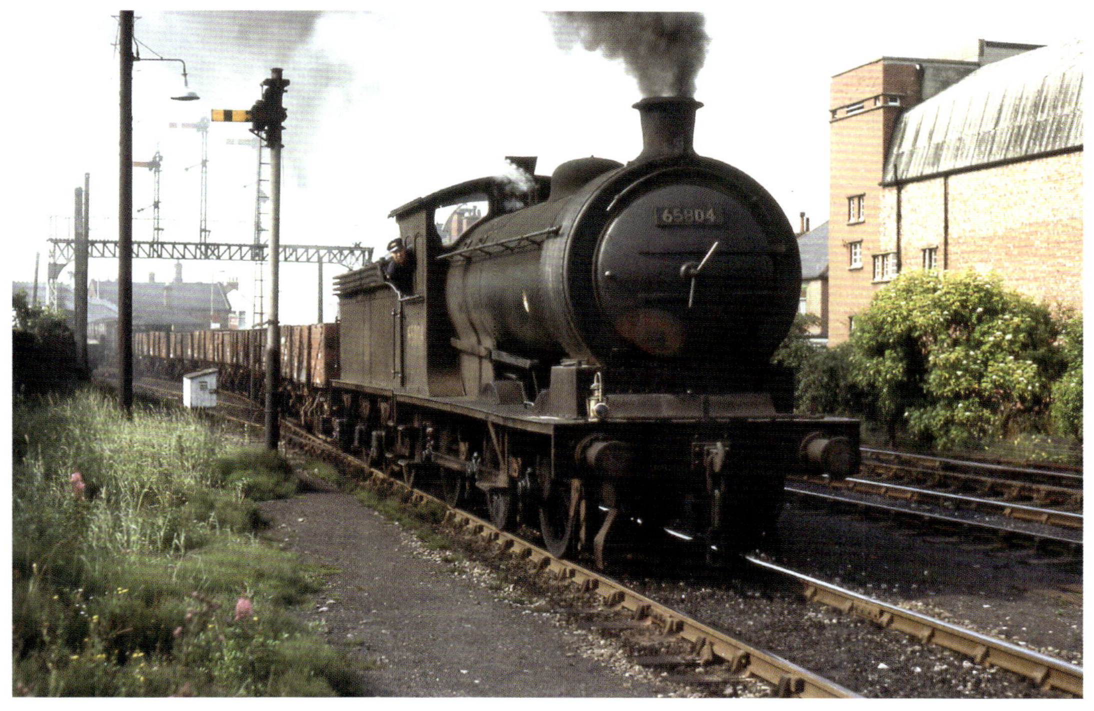

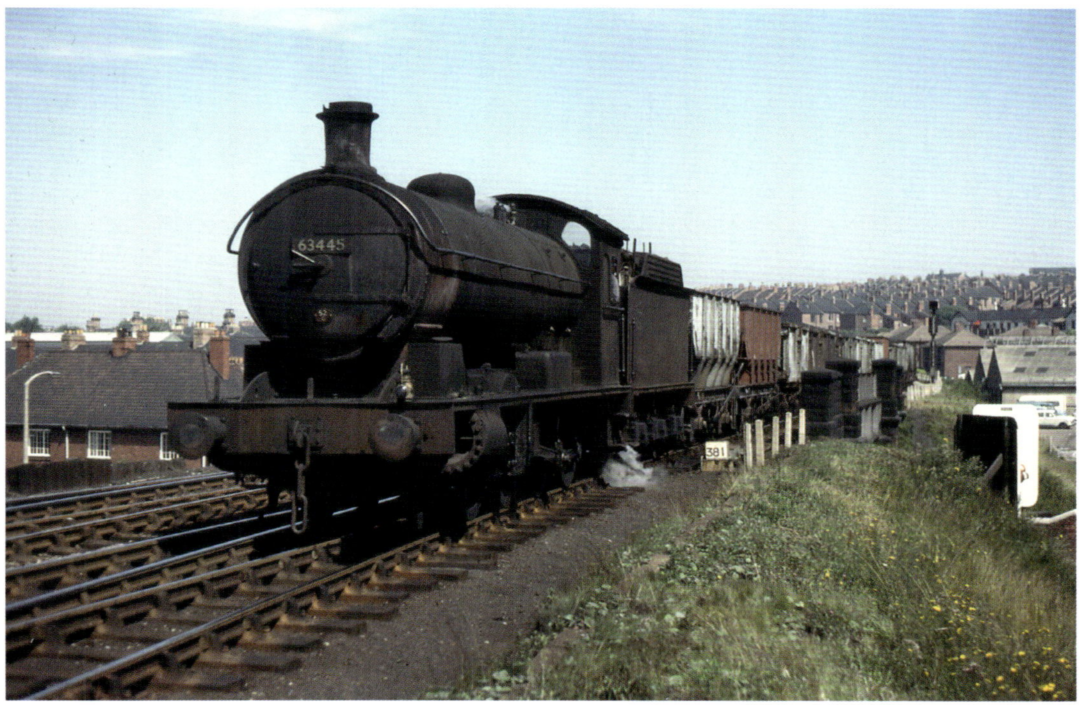

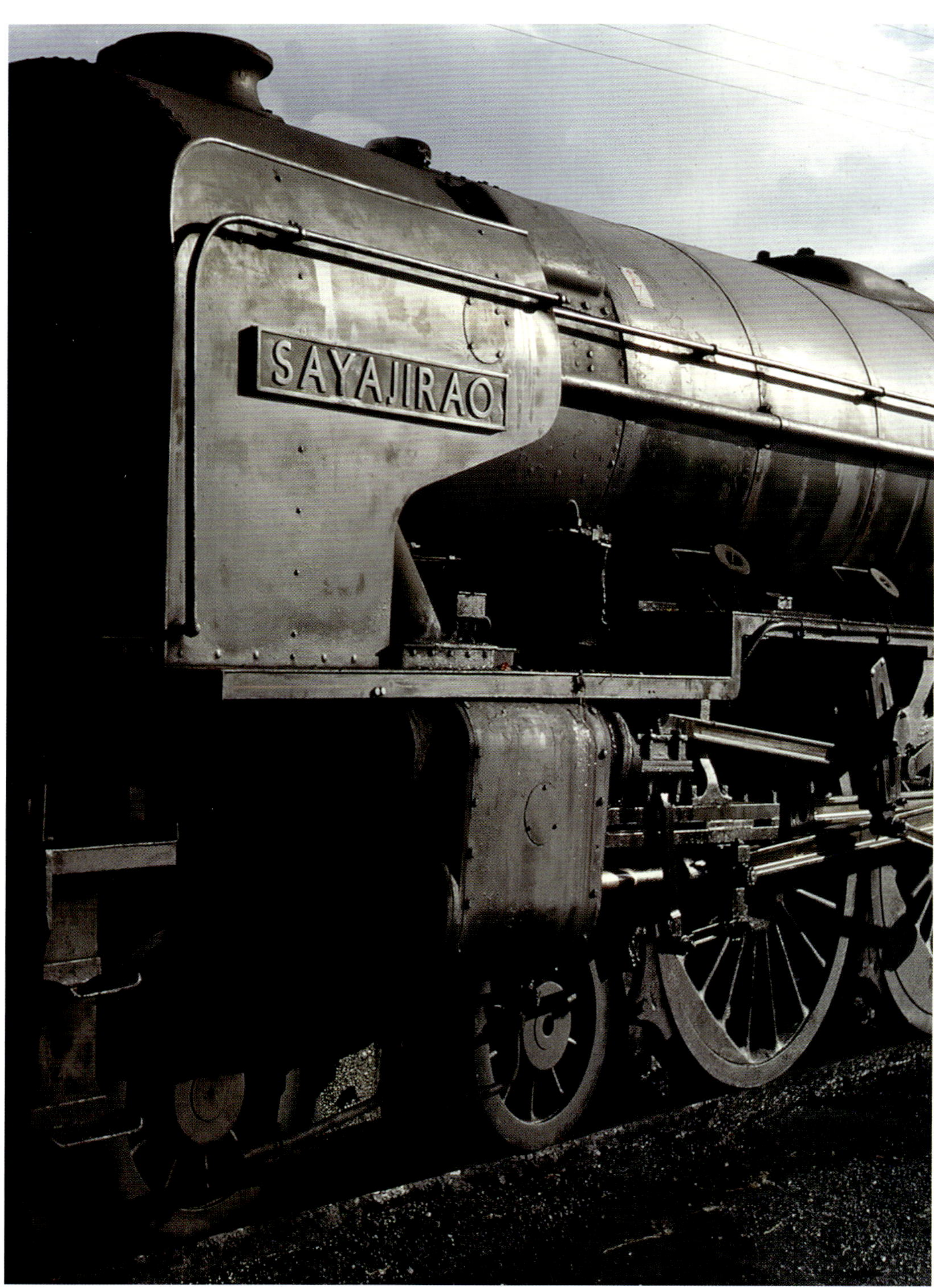

Left: Thursday 19 August 1965. At Dundee Tay Bridge shed the evening sunshine is highlighting detail at the front end of the driver's side of Class A2 4-6-2 No. 60530 *Sayajirao*. Constructed at Doncaster Works during 1948, she would be withdrawn late in 1966. In the early morning of Thursday 17 December 1953, this locomotive was derailed, at speed, while in charge of a parcels train from Edinburgh Waverley to Newcastle. Hitting debris on the track near Longniddry Junction, the fireman was killed, and the driver and guard seriously injured. (Andrew G. Forsyth)

Opposite top: September 1965. With the subsidiary signal in the 'off' position and with an excess of steam, ex-NER Class P3 (LNER Class J27) 0-6-0 No. 65879 is ready to haul a train of full coal hopper wagons into the yard at South Blyth. A product of Darlington Works in 1922, she would be withdrawn from service during 1967. (Anthony Scarsbrook)

Opposite bottom: Sunday 19 September 1965. This day saw the running of the RCTS 'Blyth & Tyne Rail Tour', which commenced at Leeds. The Newcastle to Morpeth, Blaydon and Pelaw section was in the hands of Class 4MT 2-6-0 No. 43057, which is seen here passing Choppington Crossing en route to Morpeth. Constructed post-nationalisation in 1950 at Doncaster Works, she would be withdrawn during 1966. (Anthony Scarsbrook)

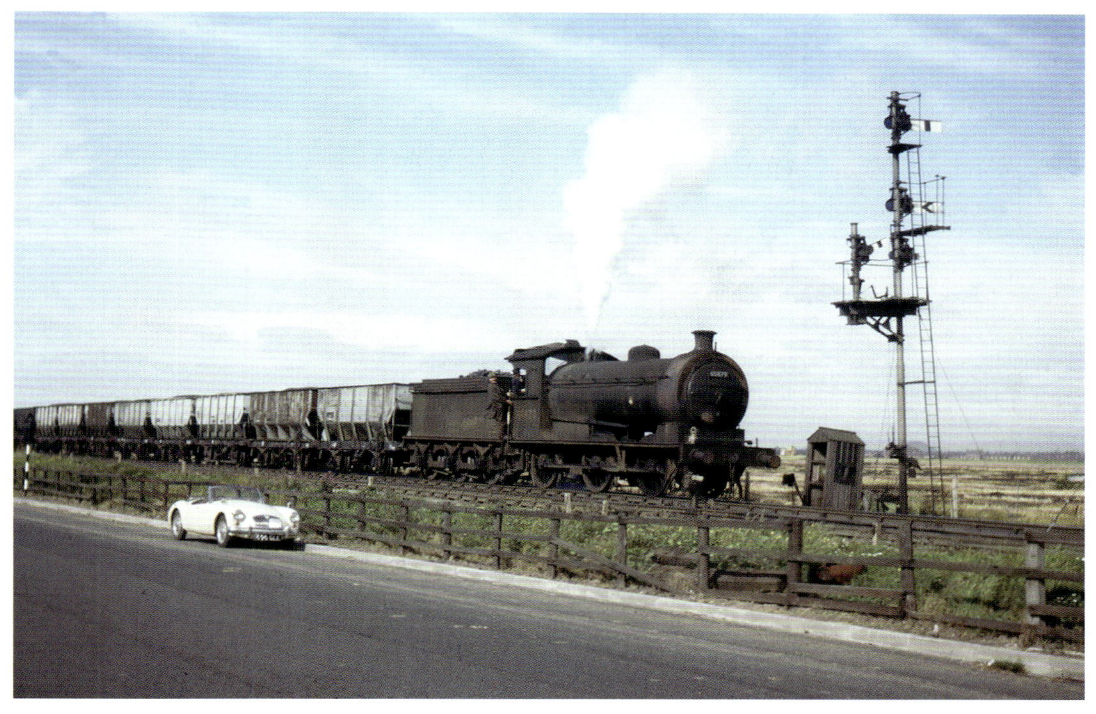

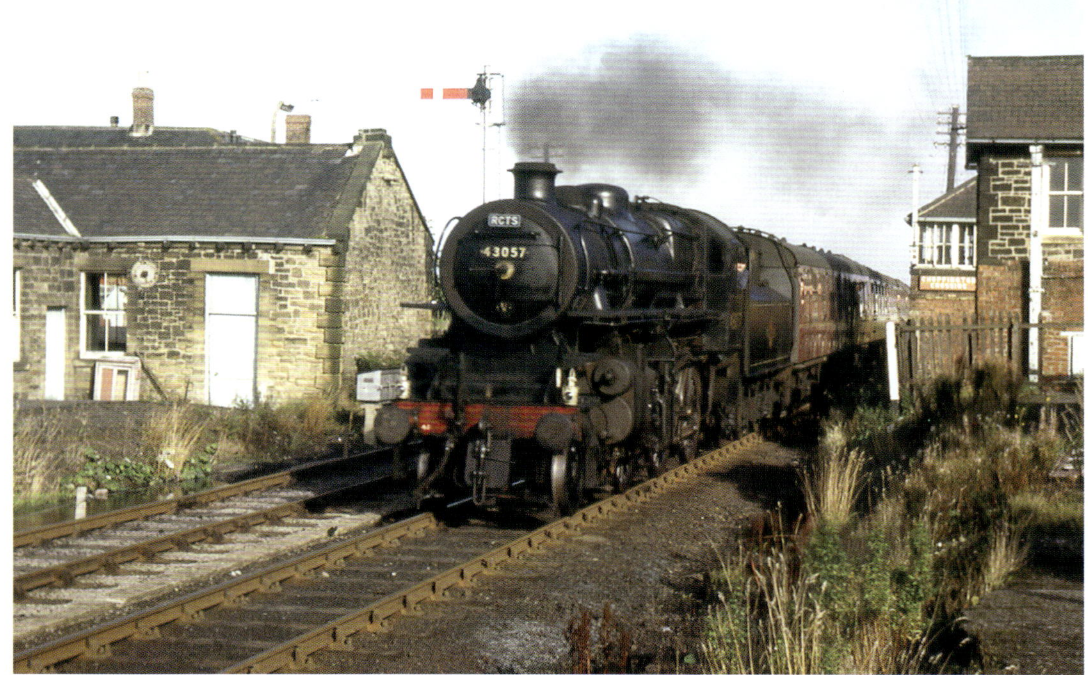

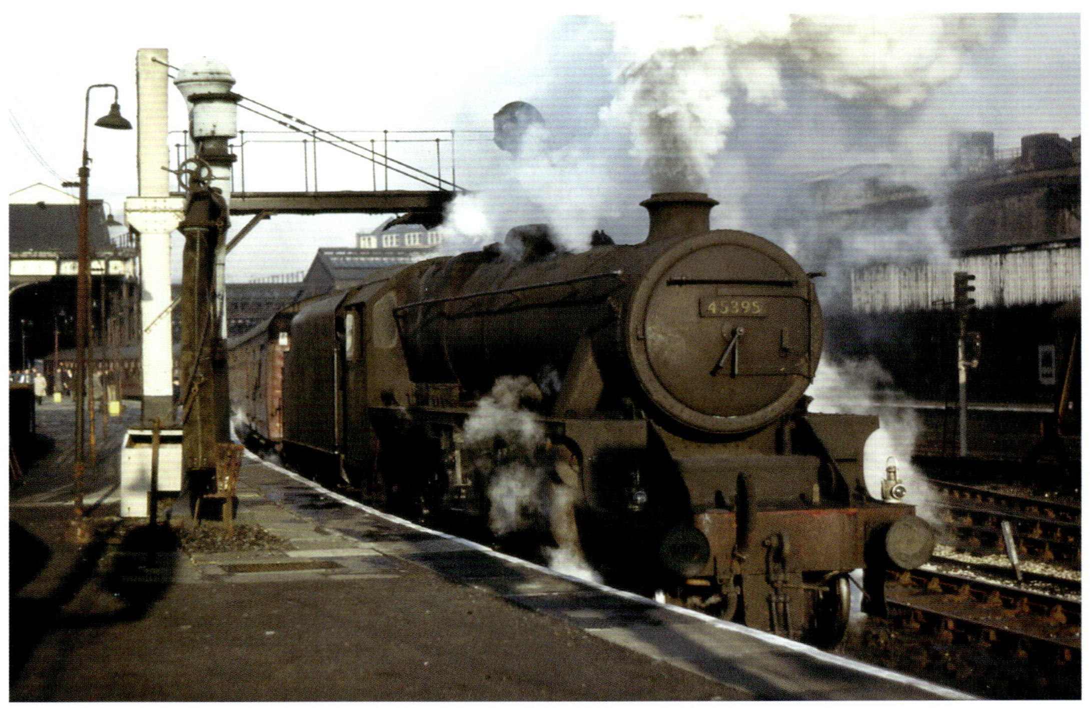

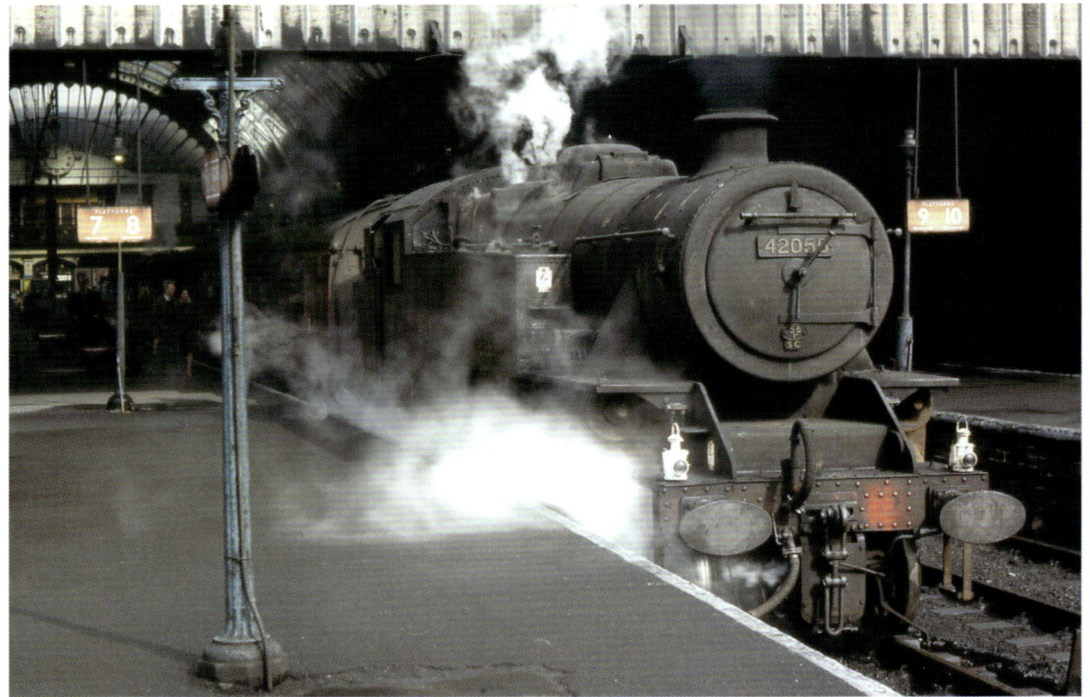

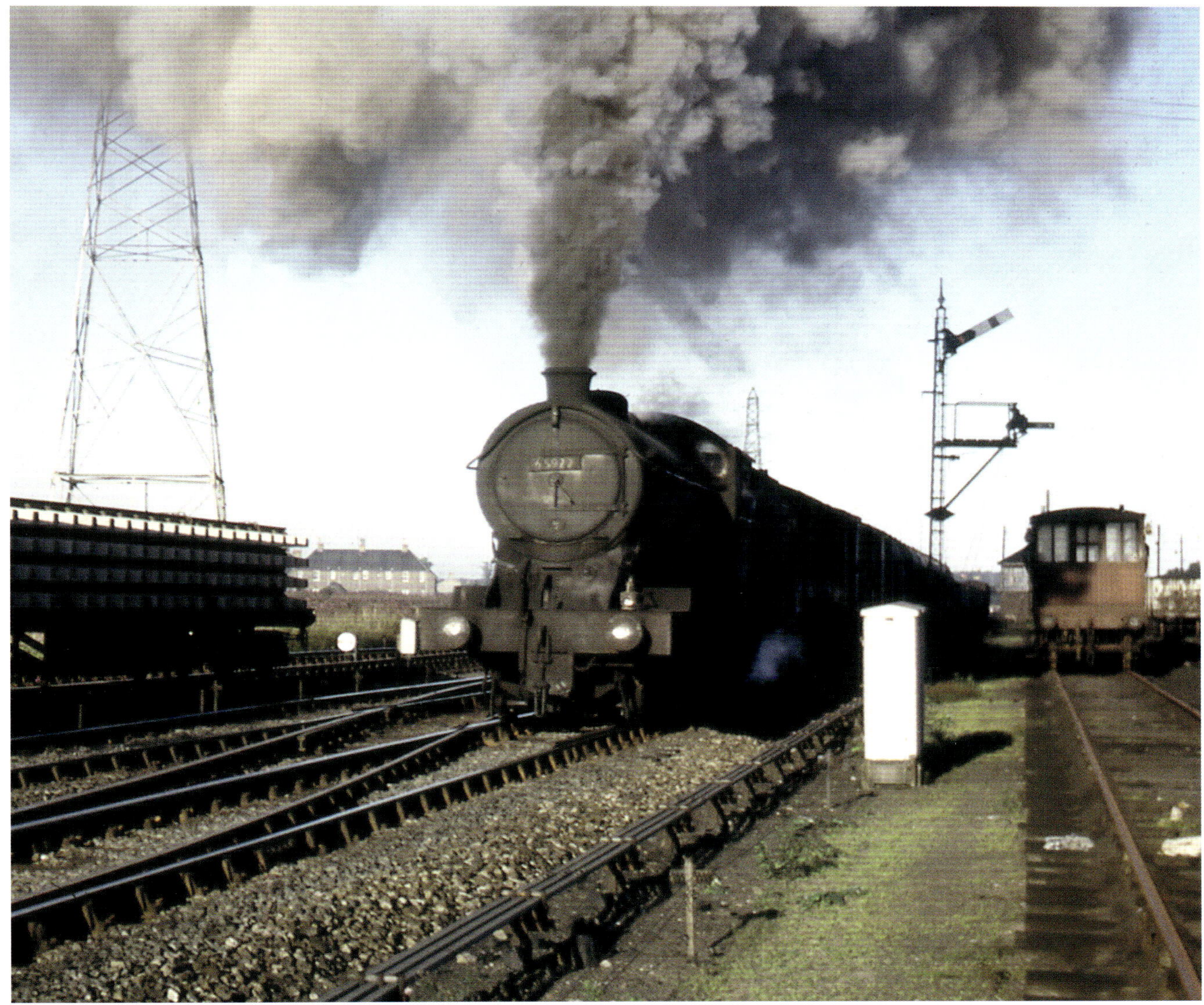

Opposite top: Thursday 24 March 1966. At Manchester Victoria station, ex-LMS Class 5 4-6-0 No. 45395 is seen departing at the head of the 5.33 p.m. working to Southport. Constructed by Armstrong Whitworth & Co. in 1937, she would be withdrawn from service during 1968. (Andrew G. Forsyth)

Opposite bottom: Saturday 26 March 1966. Waiting to depart from Bradford Exchange station with a passenger working is Class 4P 2-6-4 tank No. 42055. Entering service from Derby Works during 1950, she would be withdrawn in 1967. (Andrew G. Forsyth)

Above: August 1966. Ex-LNER Class J38 0-6-0 No. 65922 is seen working hard at the head of an 'up' goods train as it passes Thornton Junction shed. Entering service from Darlington Works during 1926 and numbered 1426 by the LNER, later becoming No. 5922, she would give forty years of service, being withdrawn two months after this photograph was taken, in October 1966. The class consisted of only thirty-five locomotives, all constructed at Darlington Works in 1926 and allocated to Scottish sheds to handle goods and mineral traffic in Fife and the central belt of Scotland. (Anthony Scarsbrook)

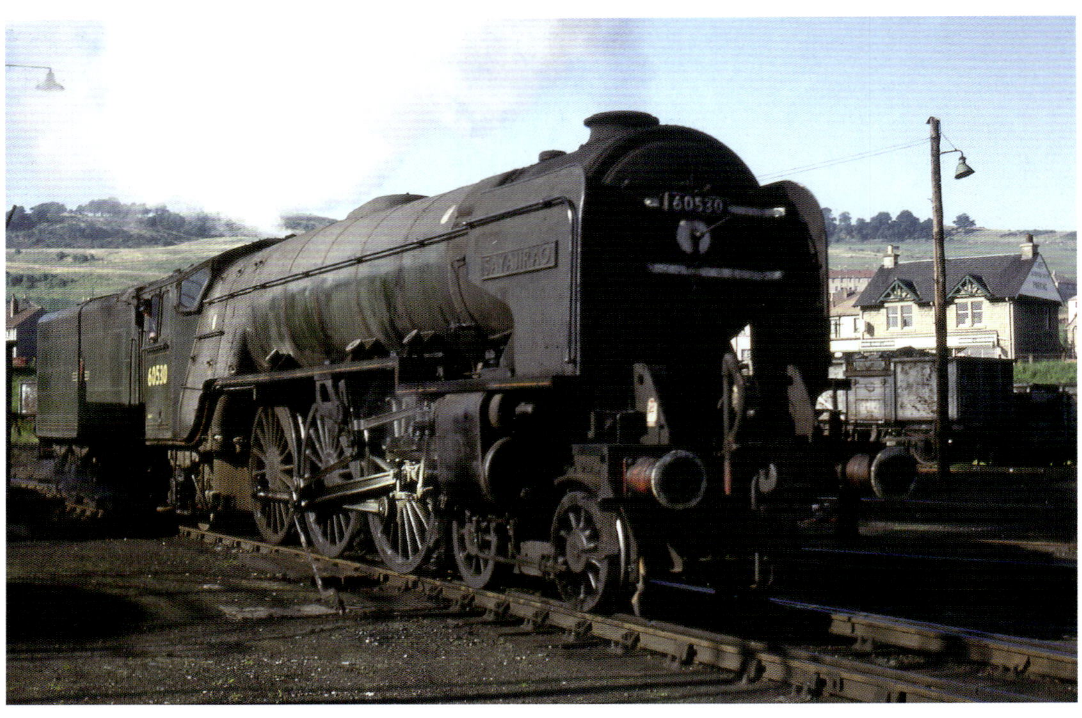

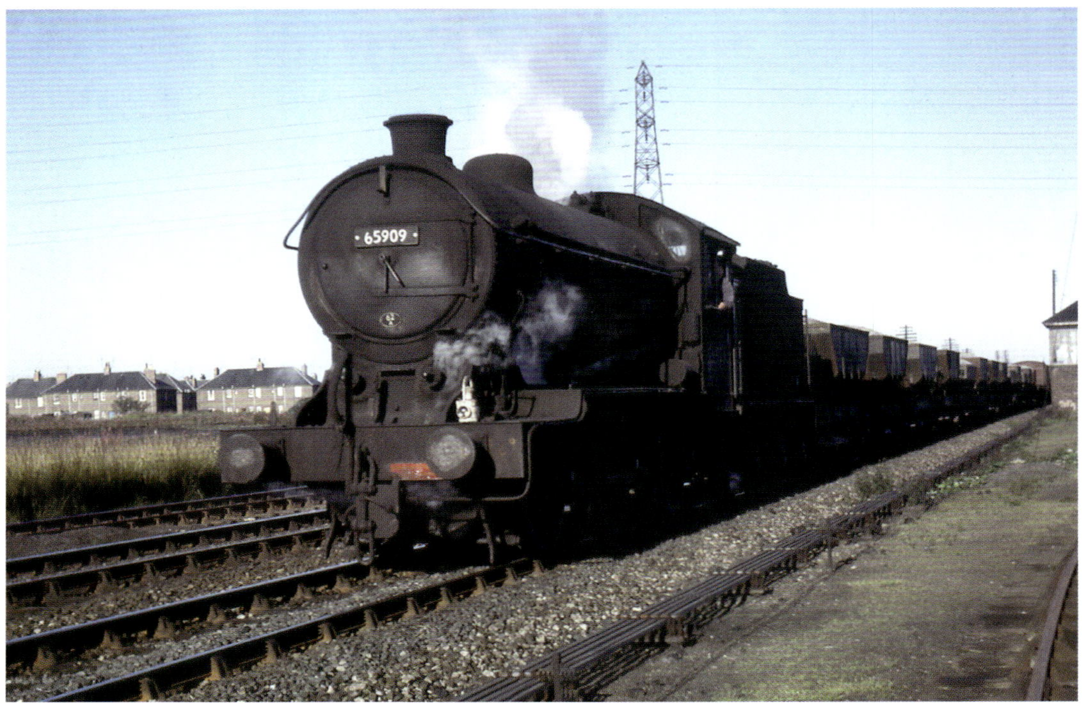

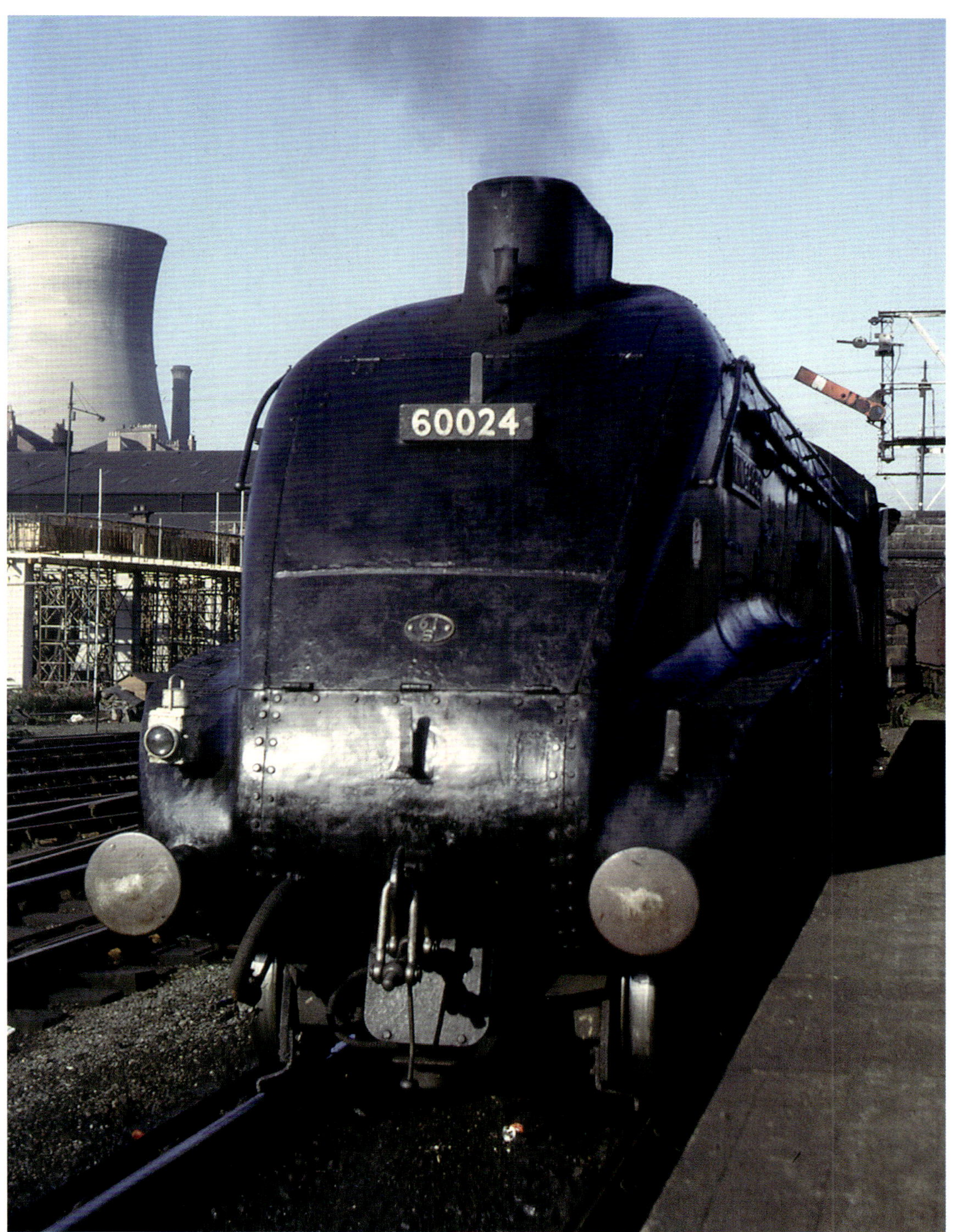

Opposite top: August 1966. Manoeuvring in Perth shed yard is Class A2 4-6-2 No. 60530 *Sayajirao* in the generally filthy condition of the period. Nearing the end of her working life, she would be withdrawn three months later, in November 1966. (Anthony Scarsbrook)

Opposite bottom: August 1966. Another goods train is seen here passing Thornton Junction shed with ex-LNER Class J38 0-6-0 No. 65909 in charge. Entering service from Darlington Works in 1926, she would be withdrawn forty years later, in 1966. (Anthony Scarsbrook)

Right: August 1966. Having arrived at Buchanan Street station in Glasgow at the head of an express from Aberdeen, ex-LNER Class A4 4-6-2 No. 60024 *Kingfisher* is reversing out of the station on its way to St Rollox shed. Bearing a 61B Ferryhill shed code, the staff there kept their allocation of Class A4s in sparkling condition. Constructed at Doncaster Works in 1936, she would be withdrawn from service during the month following the taking of this photograph, September 1966. (Anthony Scarsbrook)

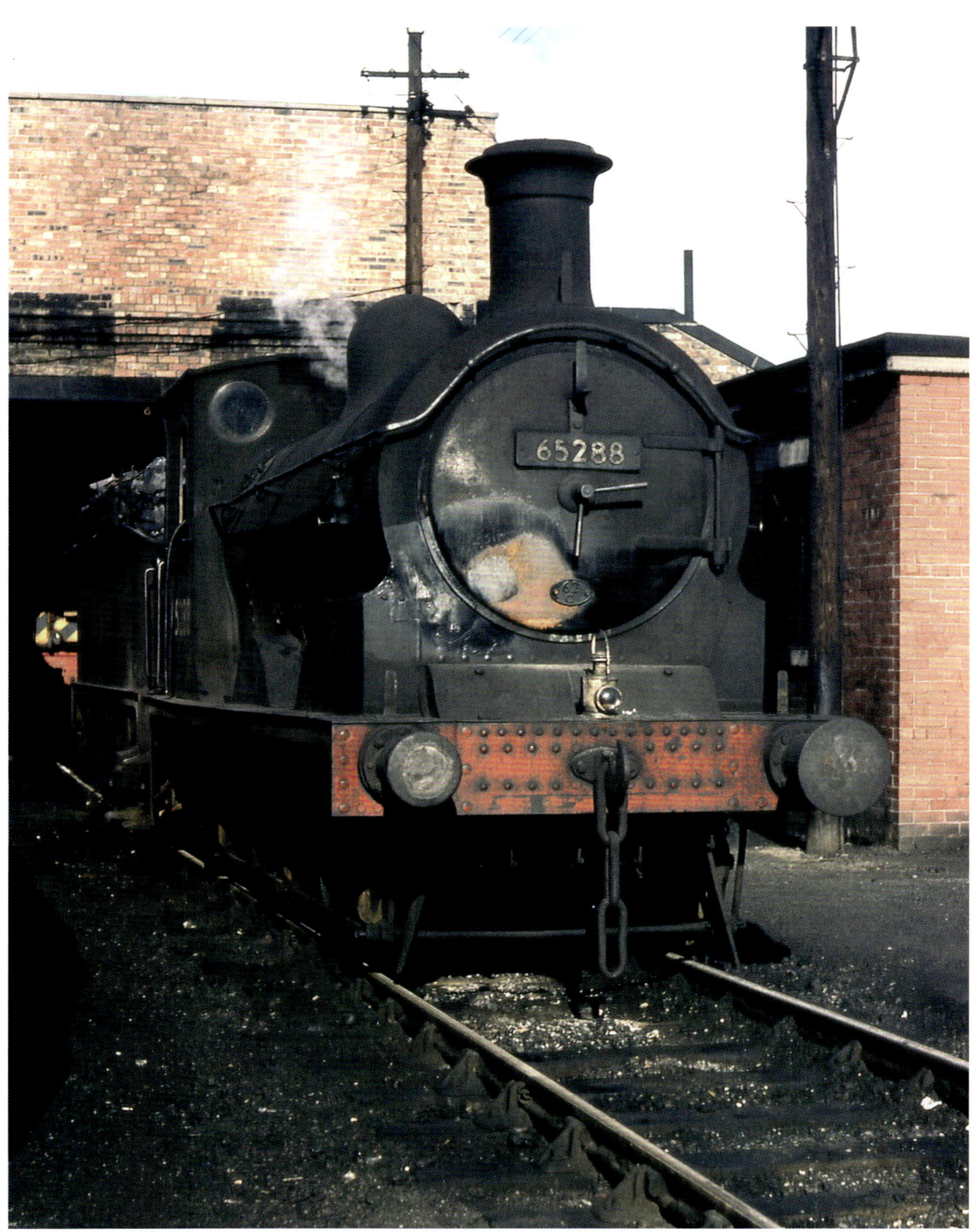

Left: Tuesday 23 August 1966. Showing signs of severe damage to her smoke-box door, ex-NBR Class C (LNER Class J36) 0-6-0 No. 65288 is quietly simmering in Dunfermline shed yard. A product of Cowlairs Works in 1897, she would be rebuilt in the form seen here during 1918 and give seventy years of service before being withdrawn during 1967. (Andrew G. Forsyth)

Opposite top: Tuesday 23 August 1966. Seen near West Ferry on the former Dundee and Arbroath Joint Line is ex-LNER Class B1 4-6-0 No. 61102 at the head of a short train of 16-ton mineral wagons destined for Carnoustie. Entering service from the NBL in 1946 and allocated new to Dundee Tay Bridge shed, she would spend her entire working life based there and be withdrawn during 1967. (Andrew G. Forsyth)

Opposite bottom: Wednesday 24 August 1966. Standing in Perth shed yard is ex-LMS Class 5 4-6-0 No. 44997 in reasonably clean condition. Constructed at Horwich Works during 1947, she would be allocated to Perth shed in 1948 and withdrawn in 1967 while based there. (Andrew G. Forsyth)

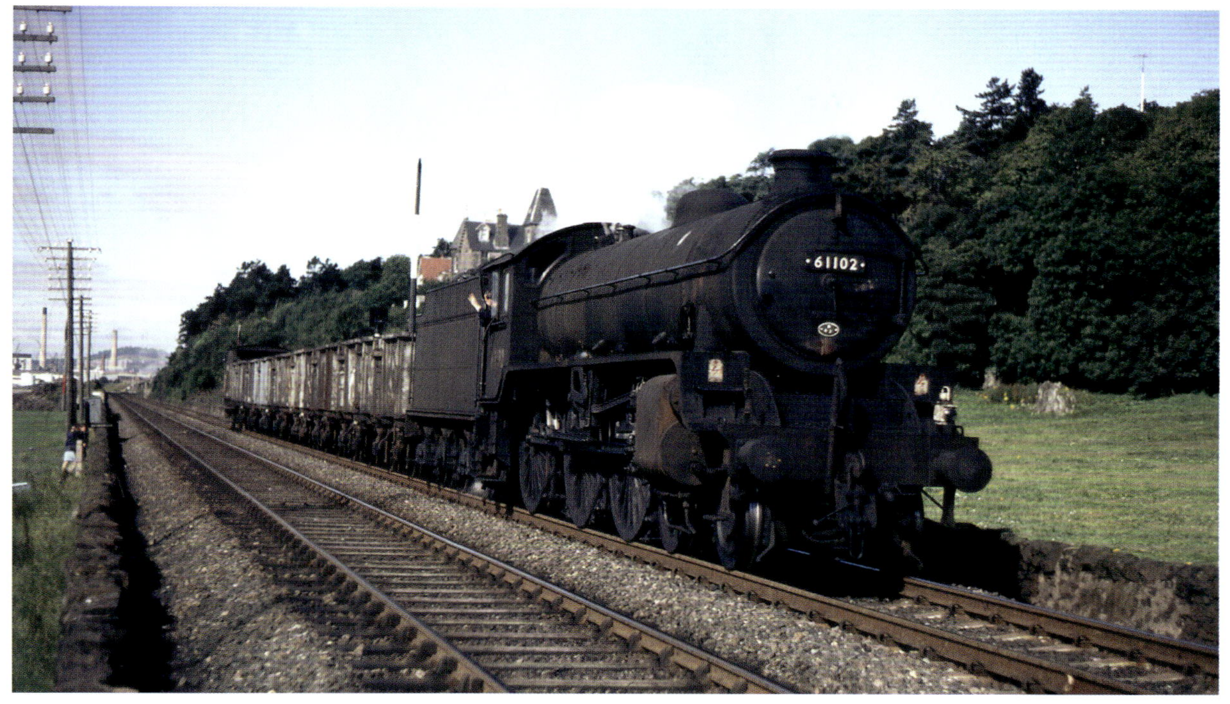

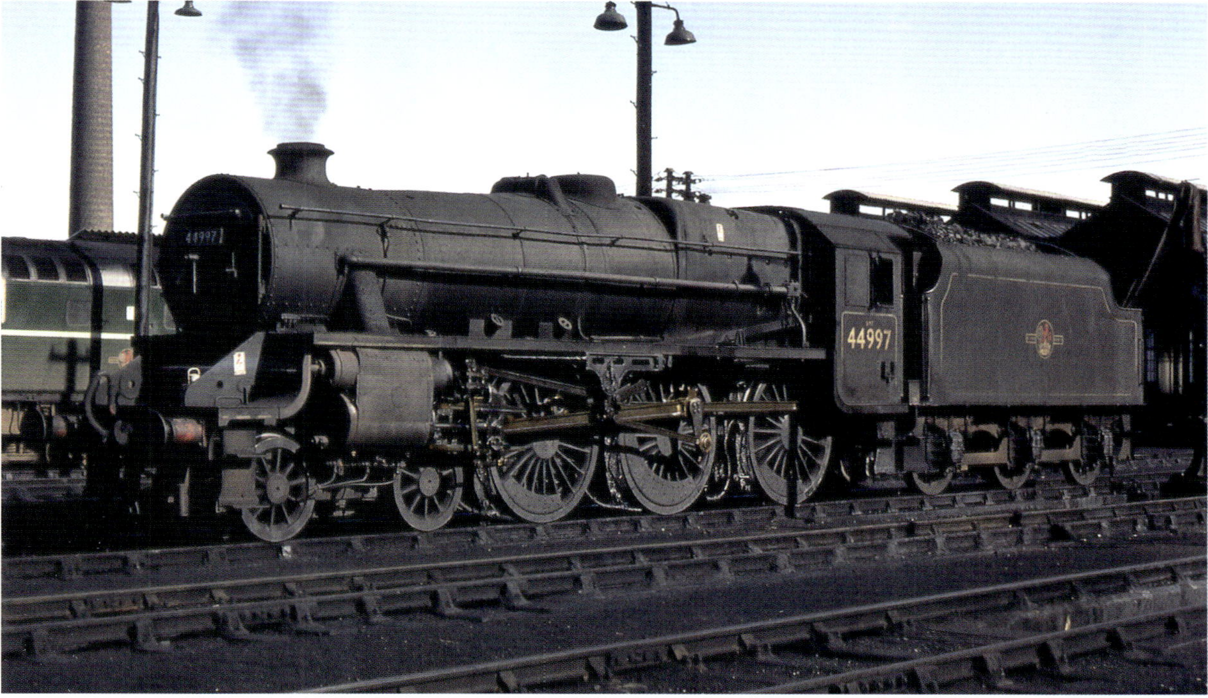

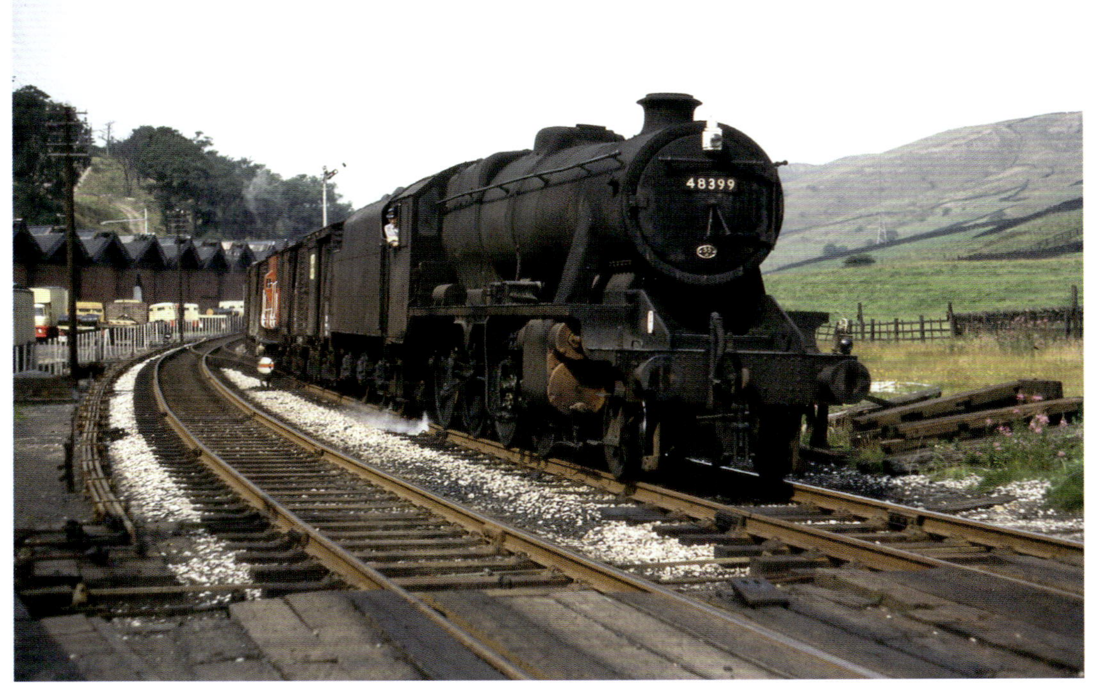

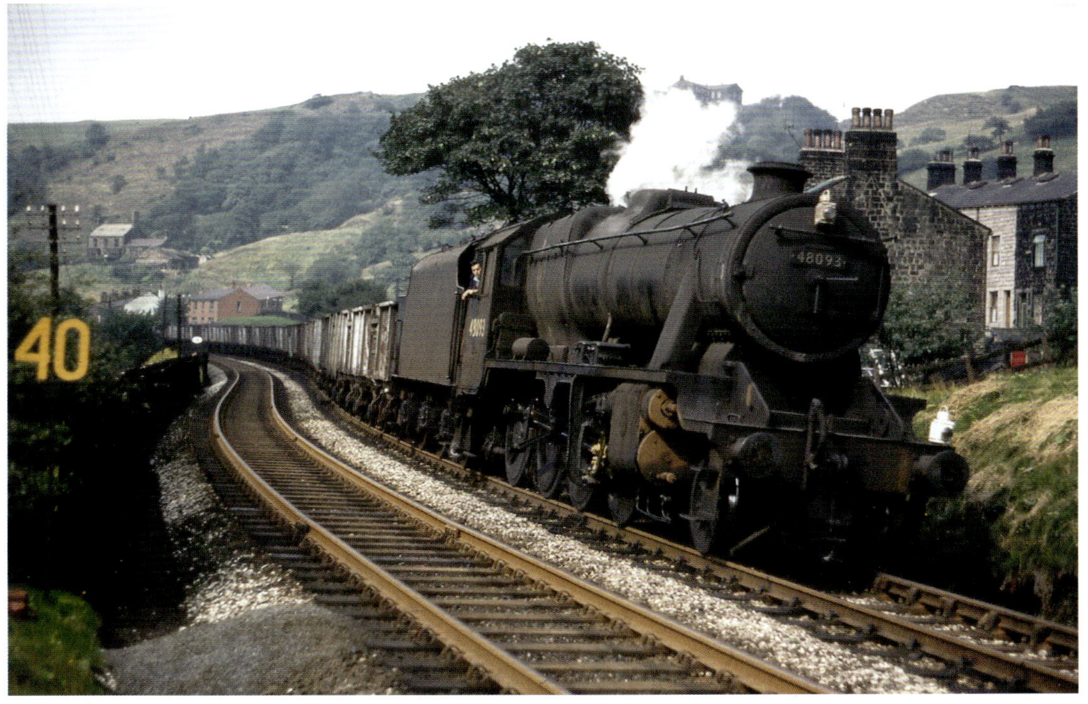

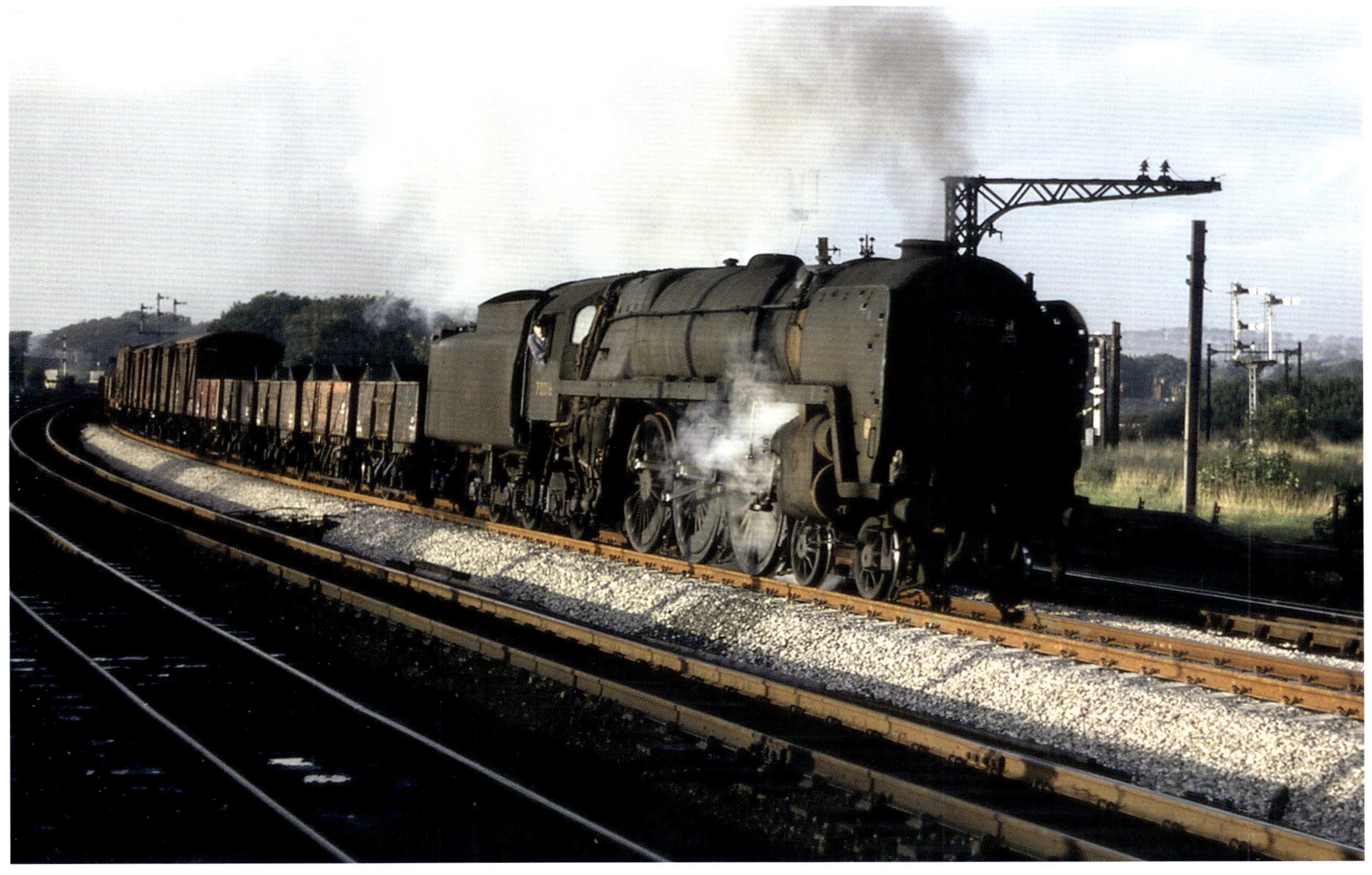

Opposite top: Wednesday 31 August 1966. At Portsmouth, on the line between Burnley and Todmorden, an 'up' goods train is in the charge of ex-LMS Class 8F 2-8-0 No. 48399. Carrying a 55A Leeds Holbeck shed code, she had been constructed at Horwich Works during 1945 and would be withdrawn from service in 1967. (Andrew G. Forsyth)

Opposite bottom: Wednesday 31 August 1966. At Cornholme, the next station in the 'up' direction from Portsmouth, another ex-LMS Class 8F 2-8-0 is leaning into the curve at the head of a train of 16-ton mineral wagons. No. 48093 was a 1937 product of the Vulcan Foundry that would give thirty years of service, before being withdrawn in 1967. (Andrew G. Forsyth)

Above: Thursday 1 September 1966. At Lancaster Castle station, BR Standard Class 7P6F Britannia 4-6-2 No. 70016, formerly named *Ariel*, is working a northbound mixed-goods train. Constructed at Crewe Works during 1951 and allocated new to Holbeck shed in Leeds, she would end her days based at Carlisle Kingmoor shed, being withdrawn in 1967. Note that her nameplates have been removed. (Andrew G. Forsyth)

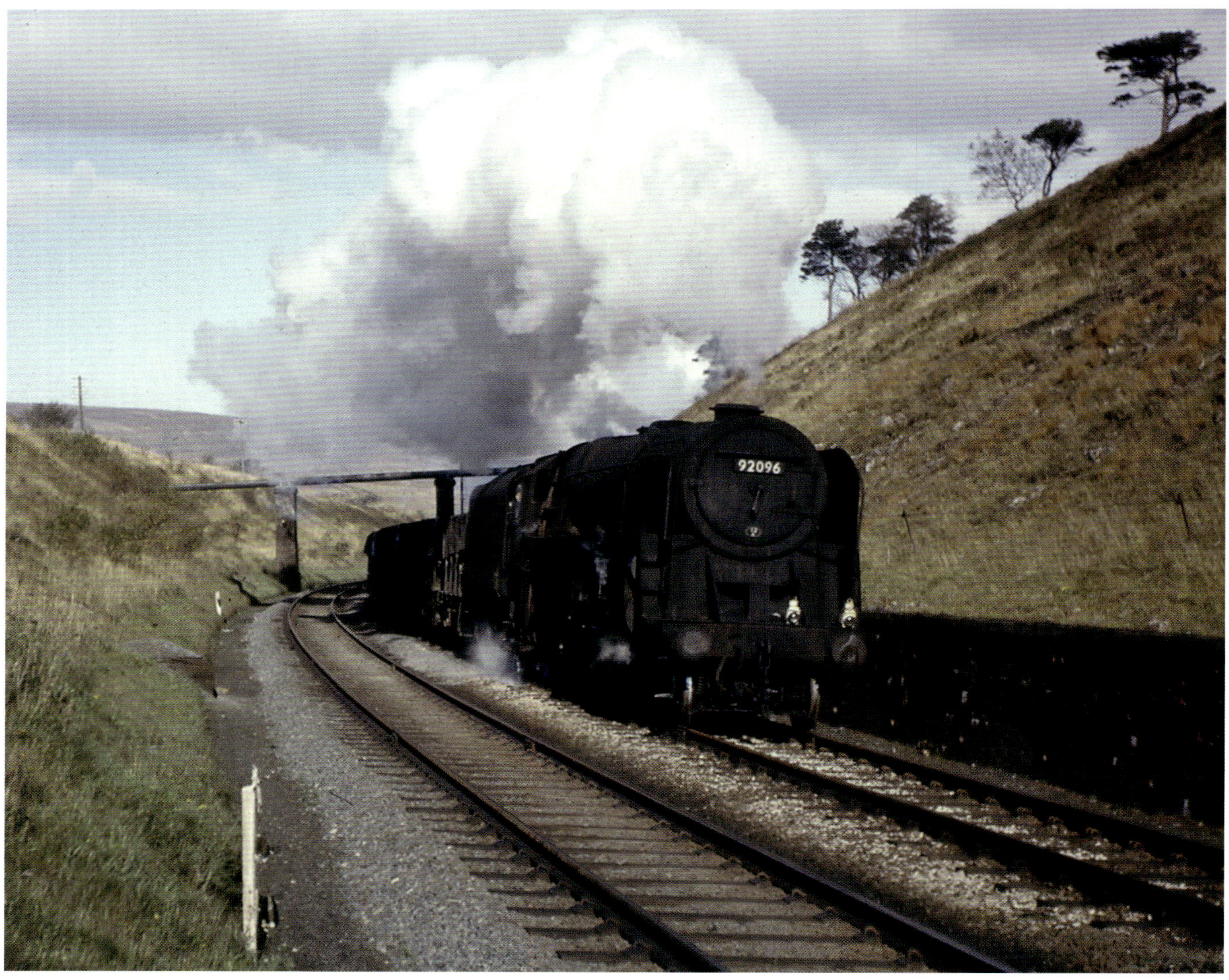

Monday 24 October 1966. Seen approaching Dent on the Settle and Carlisle route at the head of an 'up' goods working is BR Standard Class 9F 2-10-0 No. 92096. Carrying a 12A Carlisle Kingmoor shed code, she entered service from Swindon Works during 1957 and would be allocated new to Annesley shed. She would end her days at Kingmoor, being withdrawn after a short working life of only ten years, in 1967. (Anthony Scarsbrook)

The Class 9F 2-10-0s were probably the most successful of the Robert Riddles designs, being used as heavy goods locomotives throughout the Eastern, North Eastern, London Midland and Western regions of British Railways. A total of 251 examples were constructed, 198 coming from Crewe Works, with the first entering service in January 1954. A further fifty-three examples entered service from Swindon Works, the last being No. 92220 *Evening Star* as late as March 1960. Locomotives of this class could be found working at operational extremes, such as the ten examples based at Tyne Dock shed moving loads of approximately 800 tons of imported iron ore to the steel works at Consett over gradients as steep as 1 in 40. The other extreme was to see examples working passenger trains over the Somerset and Dorset section between Bath and Bournemouth, with its 70-plus miles of tortuous going, which included many severe gradients.

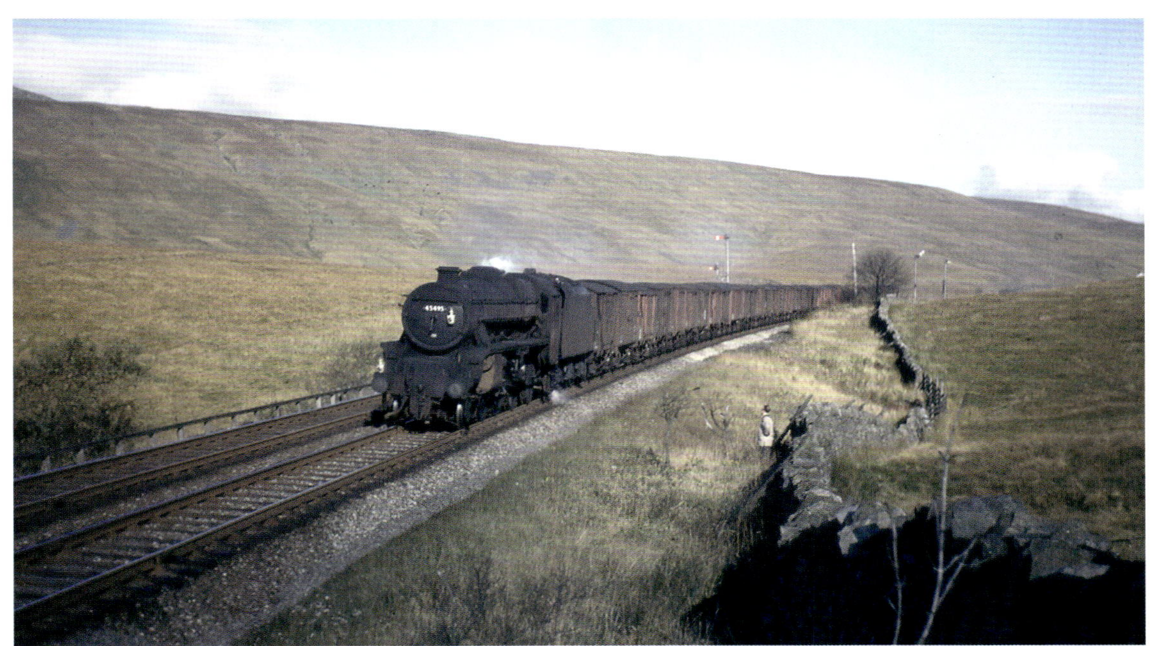

Tuesday 25 October 1966. In the open countryside between Ais Gill and Blea Moor ex-LMS Class 5 4-6-0 No. 45495 is seen at the head of an 'up' goods train. Constructed at Derby Works in 1944, she would be withdrawn during 1967. (Anthony Scarsbrook)

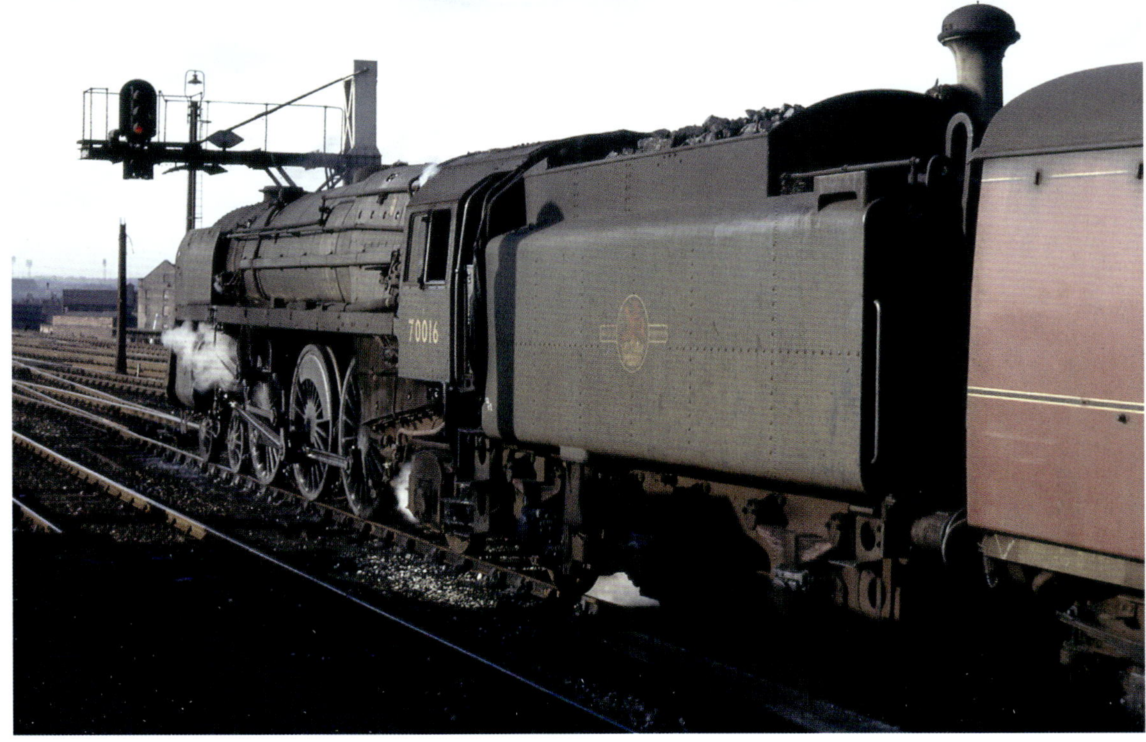

Wednesday 15 March 1967. BR Standard Class 7P6F Britannia 4-6-2 No. 70016, formerly named *Ariel*, is seen here once more, at Wigan North Western station, waiting for 'the right away' at the head of a 'down' parcels train. (Andrew G. Forsyth)

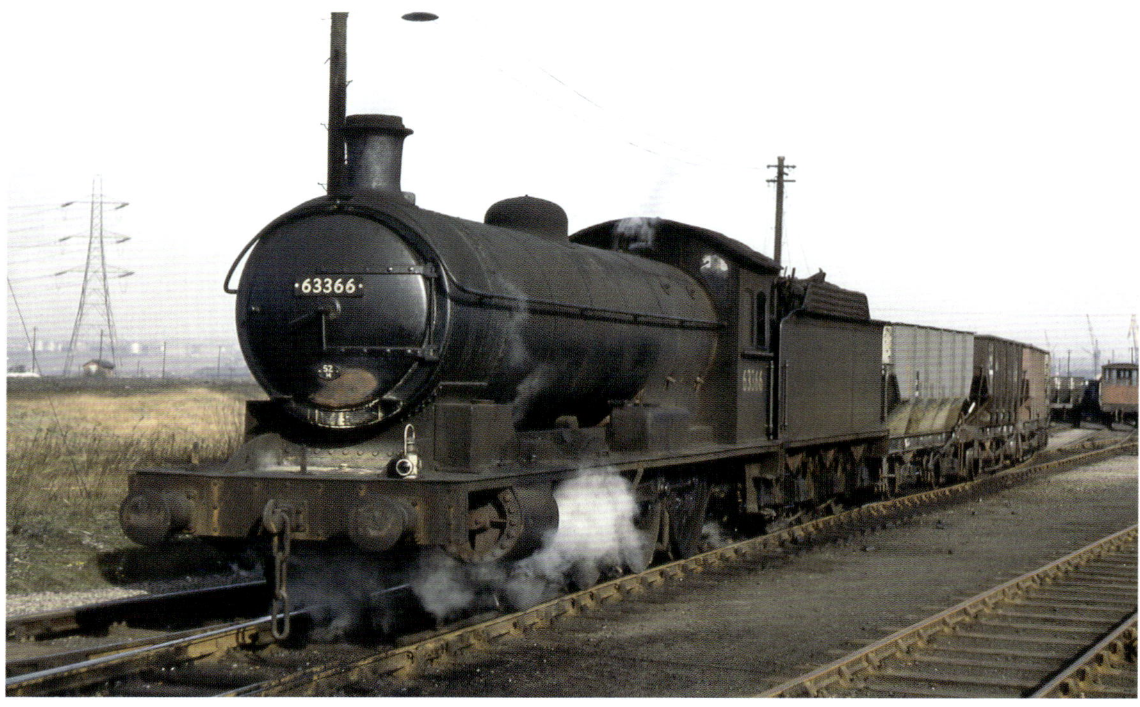

Tuesday 21 March 1967. Ex-NER Class T2 (LNER Class Q6) 0-8-0 No. 63366 carrying a 52H Tyne Dock shed code is seen departing from Tyne Dock coal sidings with a train of hopper wagons. Constructed at Darlington Works during 1913, she would be withdrawn two months after this photograph was taken. (Andrew G. Forsyth)

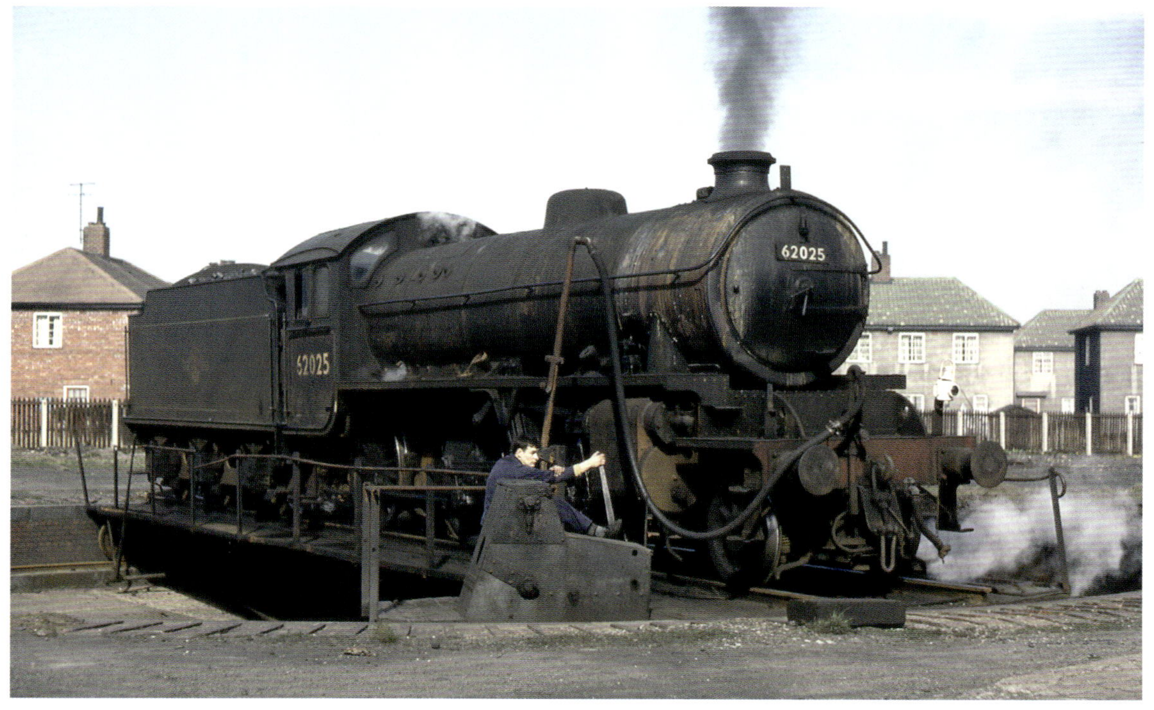

Tuesday 21 March 1967. At Tyne Dock shed, Class K1 2-6-0 No. 62025 is being turned on the vacuum-driven turntable. A product of the NBL in 1949, she would be withdrawn from service in the month following the taking of this photograph, April 1967. A total of seventy members of the class were delivered from the NBL in Glasgow over a period of ten months between May 1949 and March 1950. One of the class managed to reach the preservation scene, with No. 62005 finding its way to the North Yorkshire Moors Railway. (Andrew G. Forsyth)

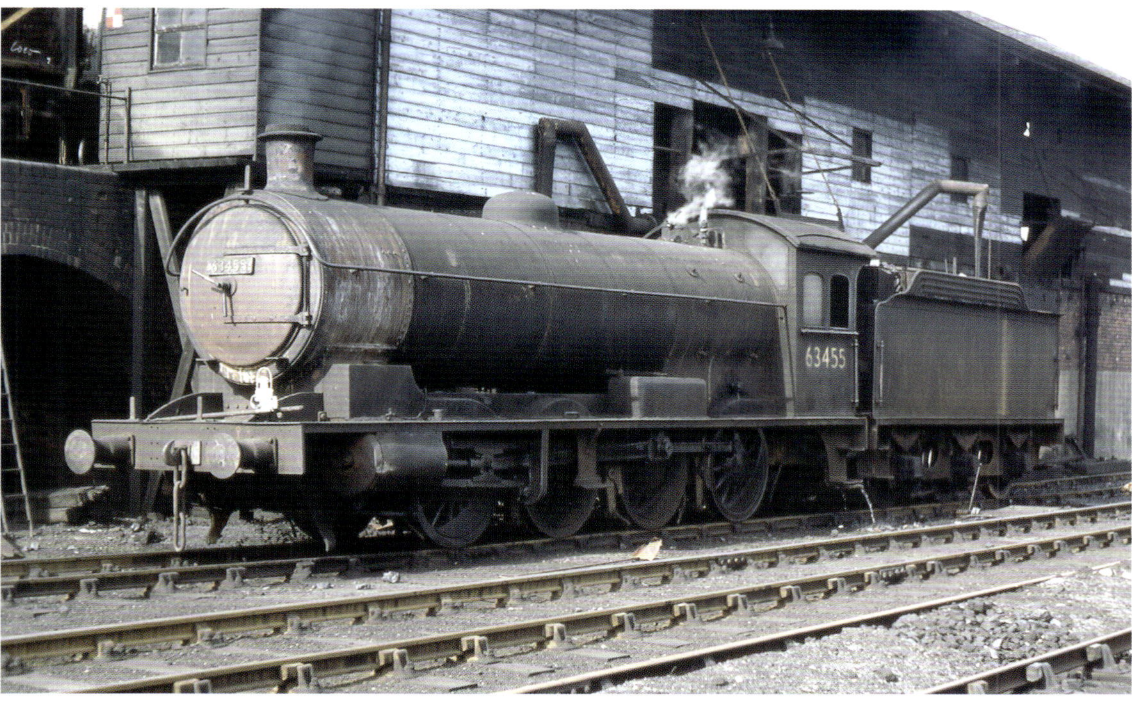

Tuesday 21 March 1967. Also at Tyne Dock shed in a deplorable condition and showing signs of serious priming is ex-NER Class T2 (LNER Class Q6) 0-8-0 No. 63455. Entering service from Armstrong Whitworth & Co. during 1920, she would be withdrawn three months after this photograph was taken, in June 1967. (Andrew G. Forsyth)

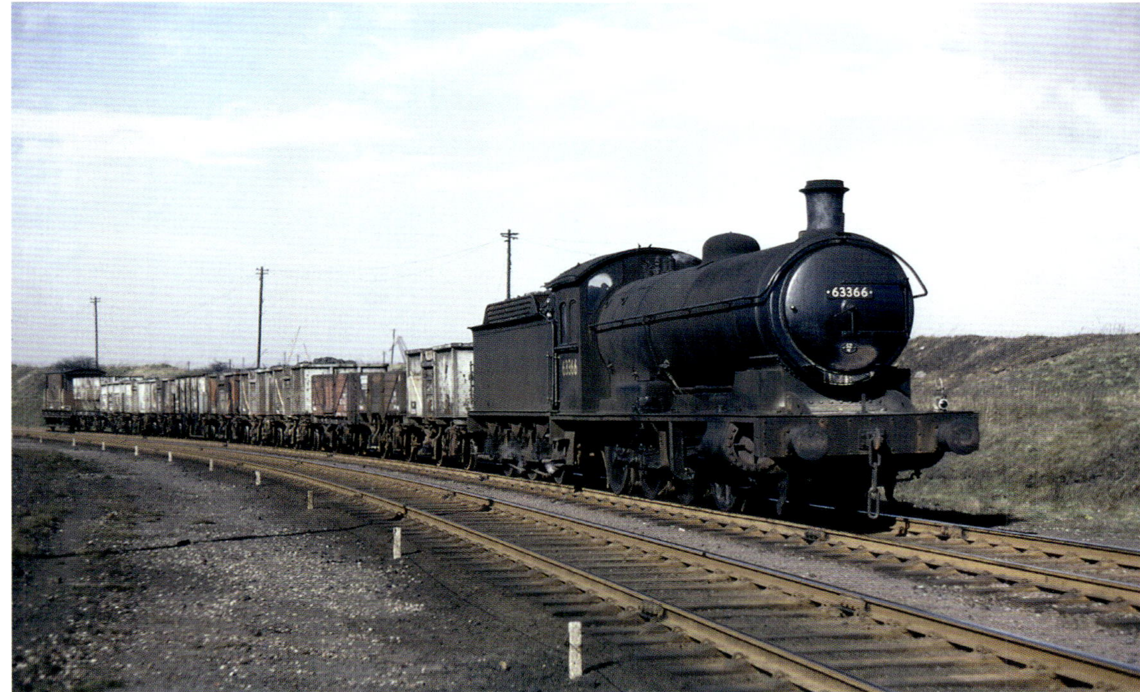

Tuesday 21 March 1967. Showing signs of hard work, Class Q6 No. 63366 is seen here again later in the day working a short train of mineral wagons near Bank Top. (Andrew G. Forsyth)

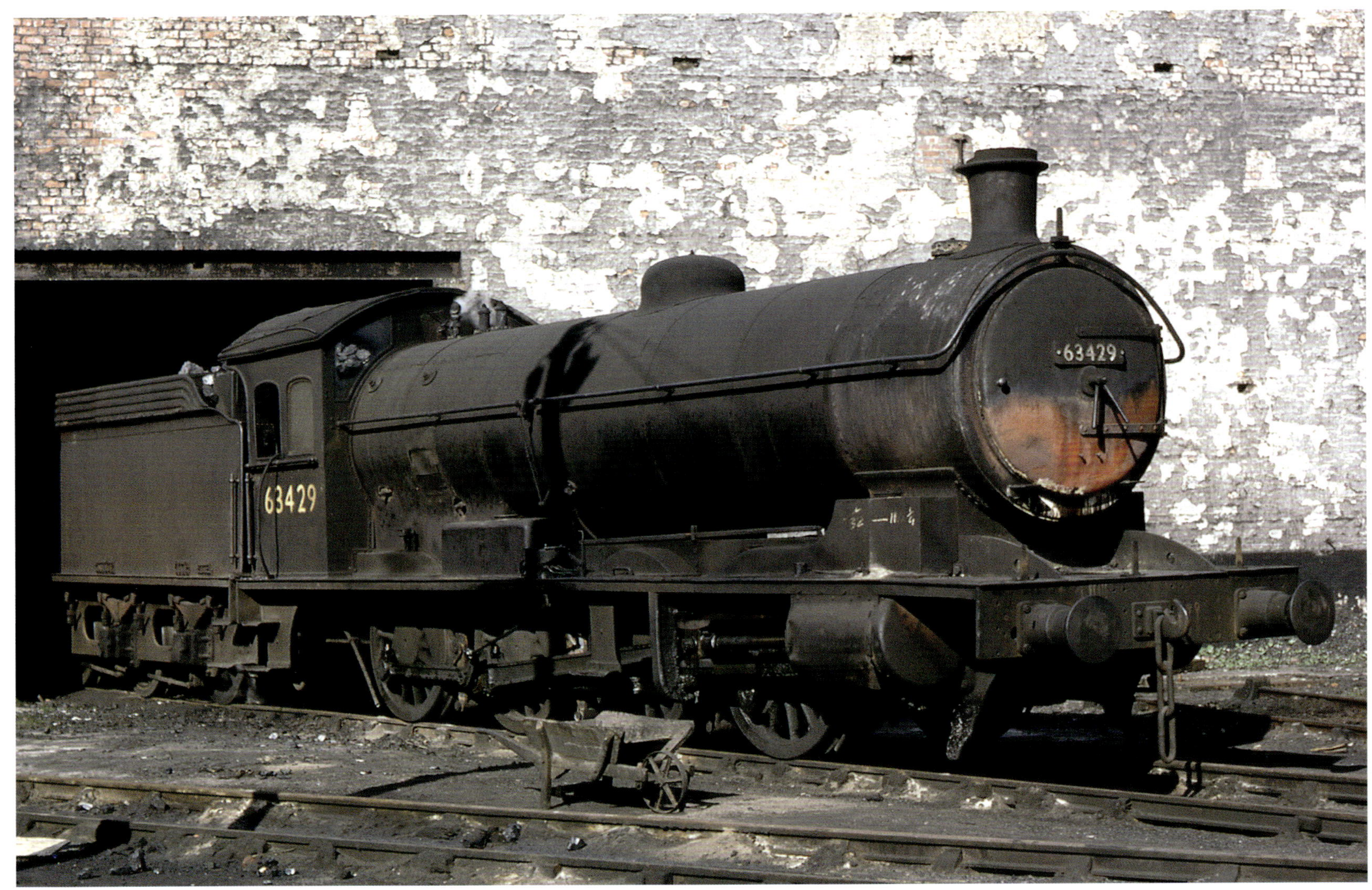

Above and opposite: Tuesday 21 March 1967. Ex-NER Class T2 (LNER Class Q6) 0-8-0 No. 63429 is seen parked in the roofless roundhouse at Tyne Dock shed. Constructed by Armstrong Whitworth & Co. during 1920 and numbered 2272 by the NER, she would become No. 3429 with the LNER and be withdrawn four months after these photographs were taken, having given forty-seven years of service. This Vincent Raven design of Class T2 was effectively a superheated boiler version of the earlier Wilson Worsdell Classes T and T1 for the NER. A total of 120 examples were constructed between 1913 and 1921, with seventy coming out of the North Eastern Railway Works at Darlington between 1913 and 1919 and a further fifty entering service from Armstrong Whitworth & Co. between 1919 and 1921. Fortunately, one example managed to reach the preservation scene, with No. 63395 being purchased by the North Eastern Locomotion Preservation Group. (Both Andrew G. Forsyth)

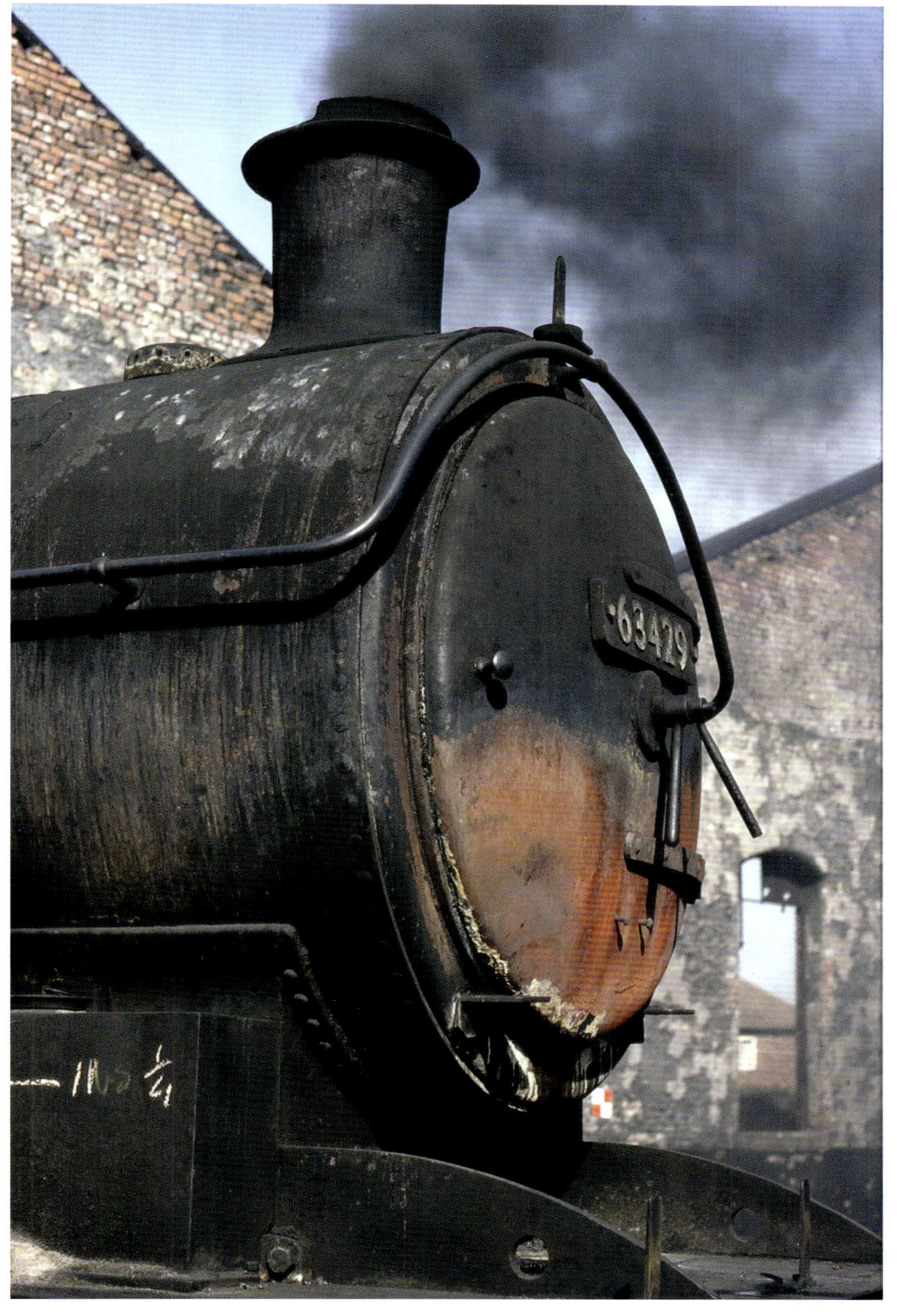

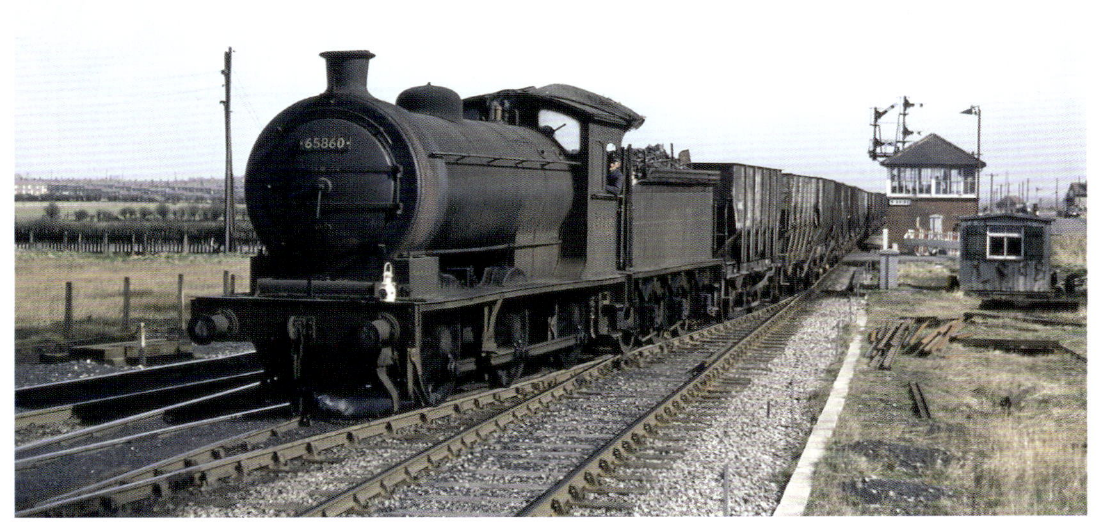

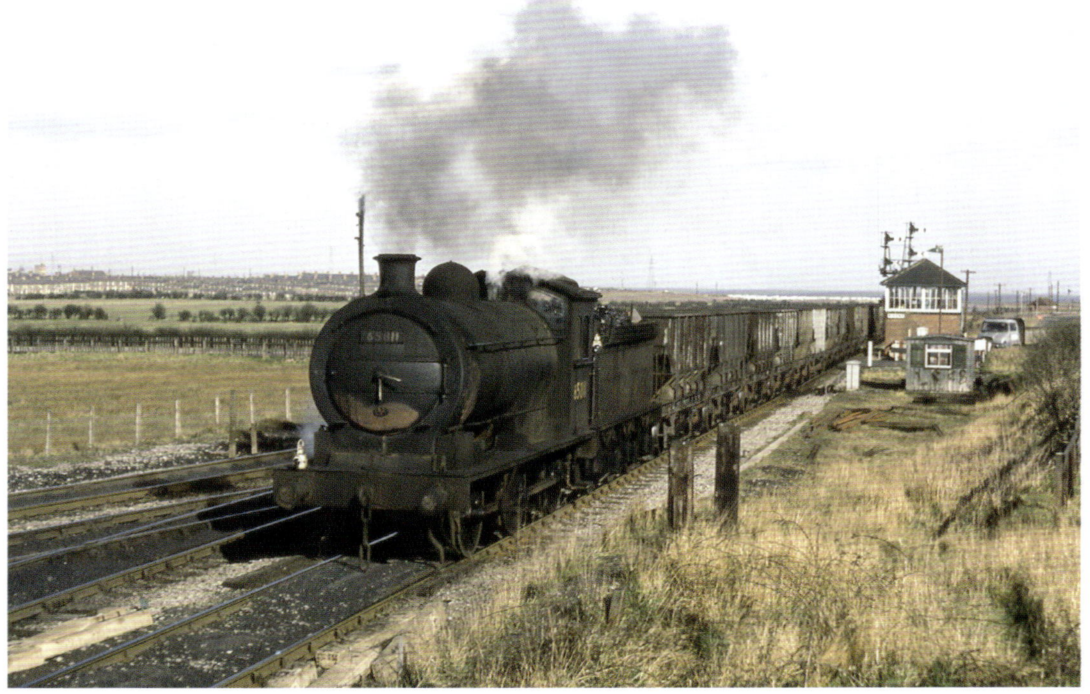

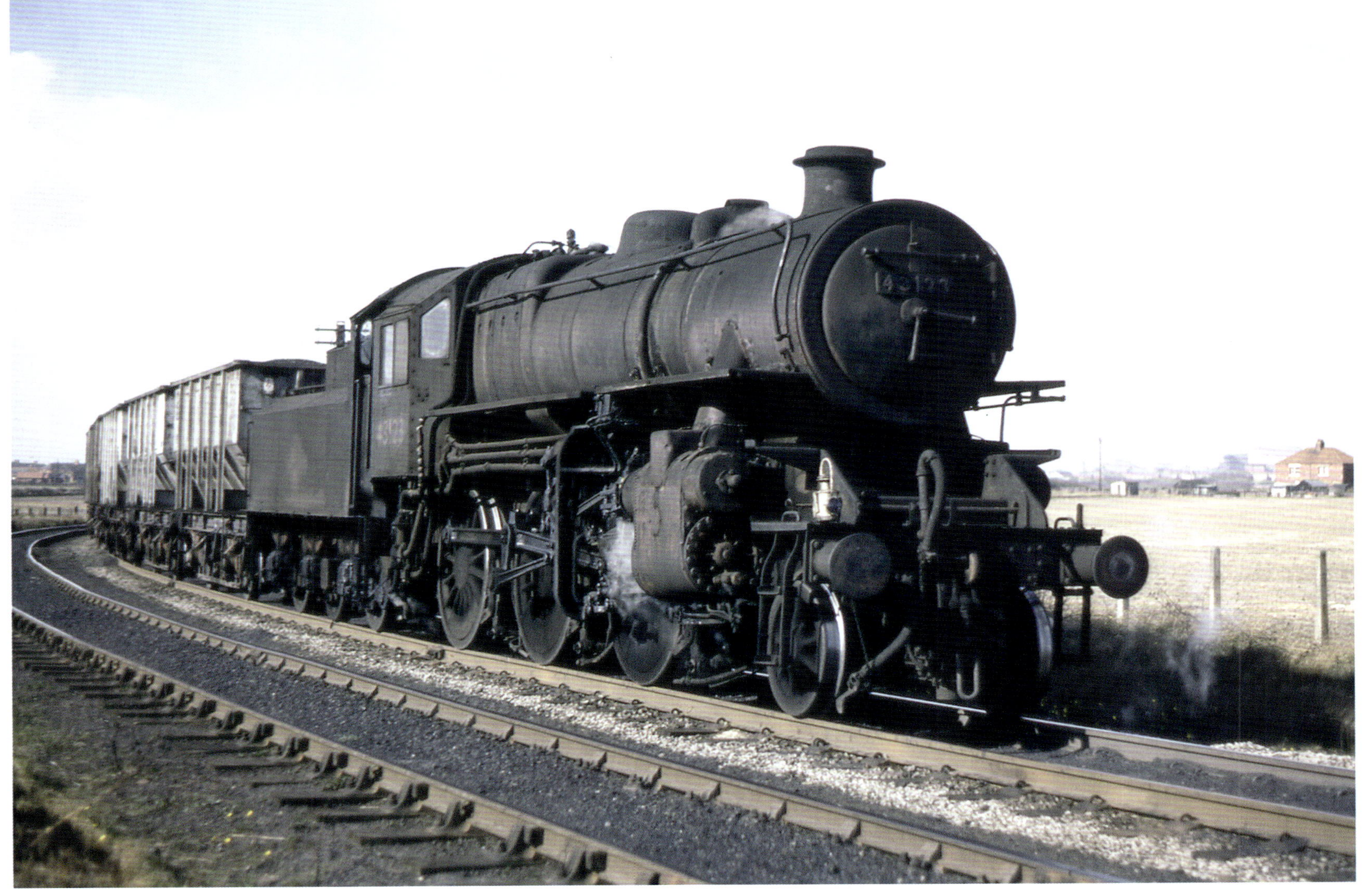

Opposite top: Wednesday 22 March 1967. Ex-NER Class P3 (LNER Class J27) 0-6-0 No. 65860 is seen passing Winning signal box at the head of an 'up' train of coal hopper wagons. Constructed at Darlington Works during 1921, she would be withdrawn from service four months after this photograph was taken, in July 1967. Winning signal box controlled the junction on the route from North Blyth to both Bedlington and Ashington, and this train is seen crossing to the Ashington line. (Andrew G. Forsyth)

Opposite bottom: Wednesday 22 March 1967. Also seen passing Winning signal box in charge of an 'up' train of loaded coal hopper wagons is ex-NER Class P3 (LNER Class J27) 0-6-0 No. 65811. A product of the NBL in 1908, she would give fifty-nine years of service before being withdrawn later in 1967. (Andrew G. Forsyth)

Above: Wednesday 22 March 1967. Class 4MT 2-6-0 No. 43123 is seen coming off the Ashington route towards Winning signal box with a 'down' train of coal hopper wagons. Entering service from Horwich Works in 1951, she would be withdrawn later during 1967. (Andrew G. Forsyth)

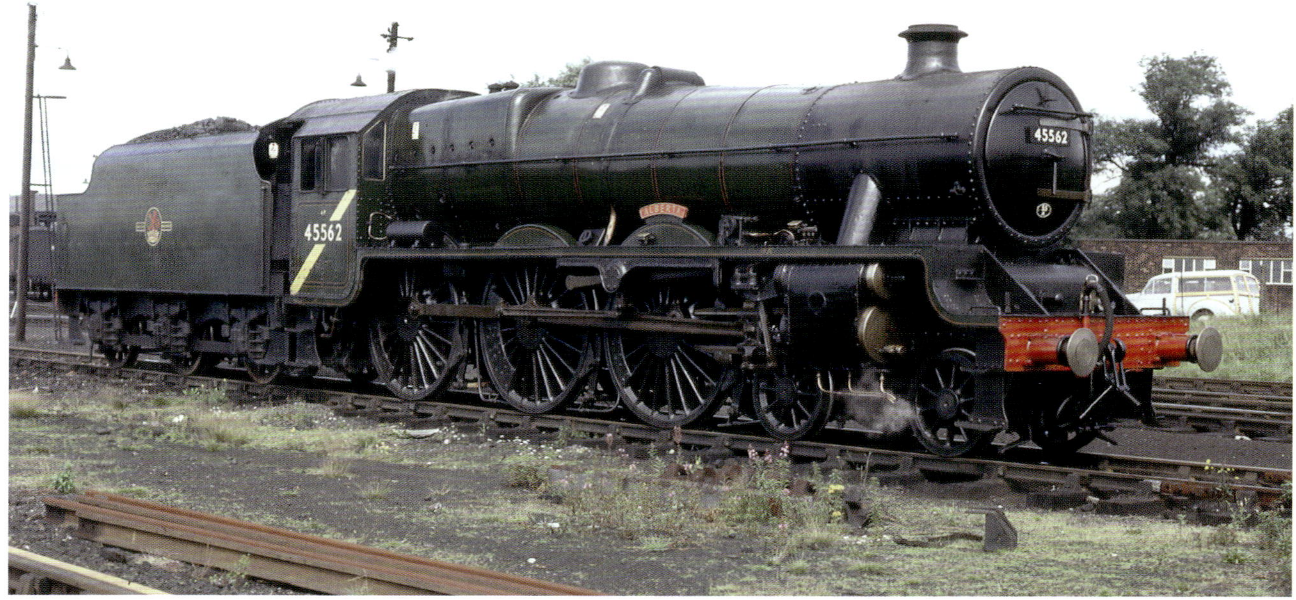

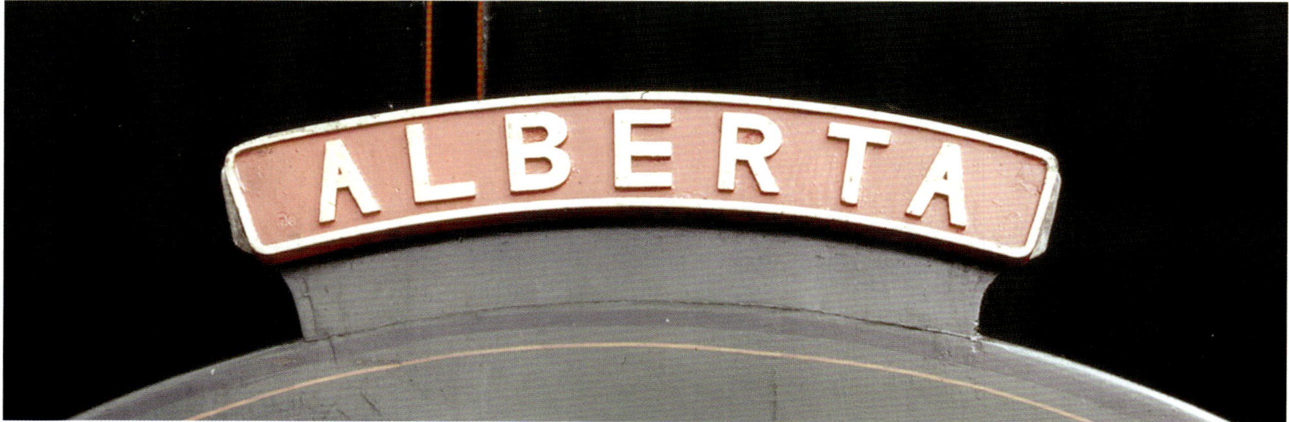

Saturday 19 August 1967. *Top:* At Kingmoor shed in Carlisle, ex-LMS Class 6P5F Jubilee 4-6-0 No. 45562 *Alberta* is in sparkling condition and is carrying the diagonal yellow stripe on her cab side that denotes she is not allowed to work under the AC overhead lines south of Crewe. Constructed by the NBL in 1934, she would be withdrawn during the first week of November 1967. *Bottom:* The fireman's side nameplate. Based at Leeds Holbeck (55A) at this time, she would be kept in immaculate condition by the shed staff there and would become the last steam locomotive to haul the Royal Train in May 1967. One of her last workings was the Manchester Railway Travel Society/Severn Valley Railway special on 28 October 1967, one week before her official withdrawal from service. (Both Andrew G. Forsyth)

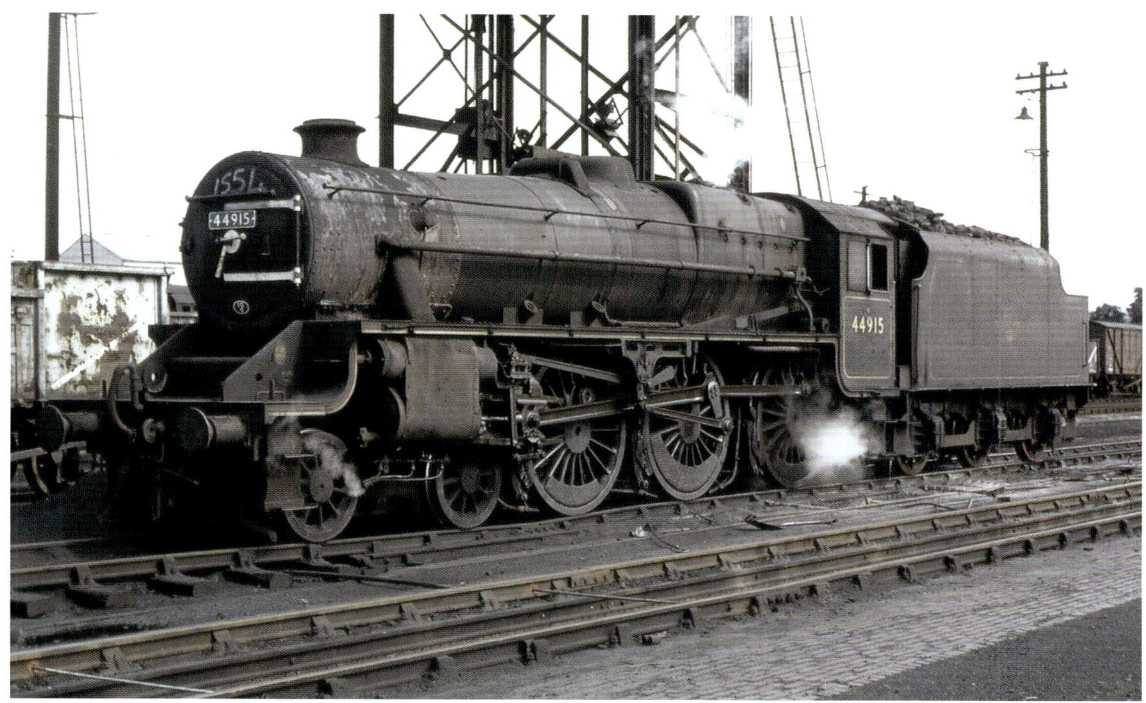

Saturday 19 August 1967. At Kingmoor shed, ex-LMS Class 5 4-6-0 No. 44915 is carrying a 10D Lostock Hall shed code. A product of Crewe Works in 1945, she would be withdrawn four months after this photograph was taken, during December 1967. (Andrew G. Forsyth)

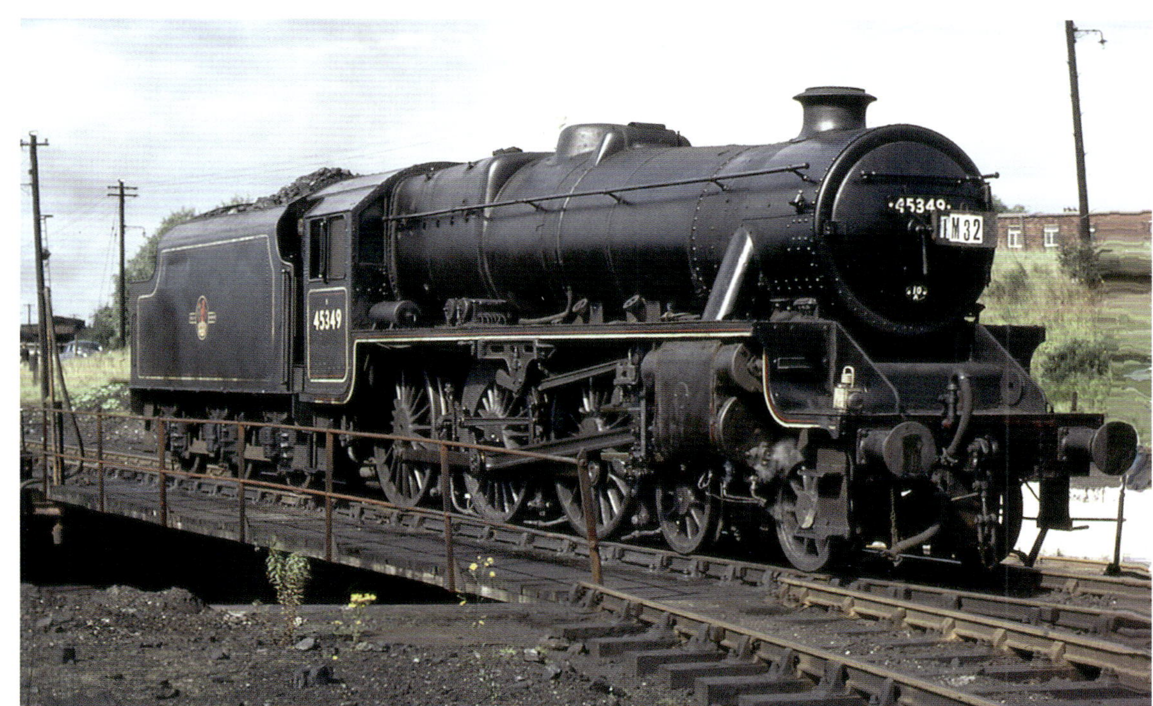

Saturday 19 August 1967. In comparison to the previous photograph, ex-LMS Class 5 4-6-0 No. 45349 has been given a good clean by the staff at Carnforth shed, whose 10A shed code she is bearing. Constructed by Armstrong Whitworth & Co. during 1937, she would be withdrawn at the end of 1967. (Andrew G. Forsyth)

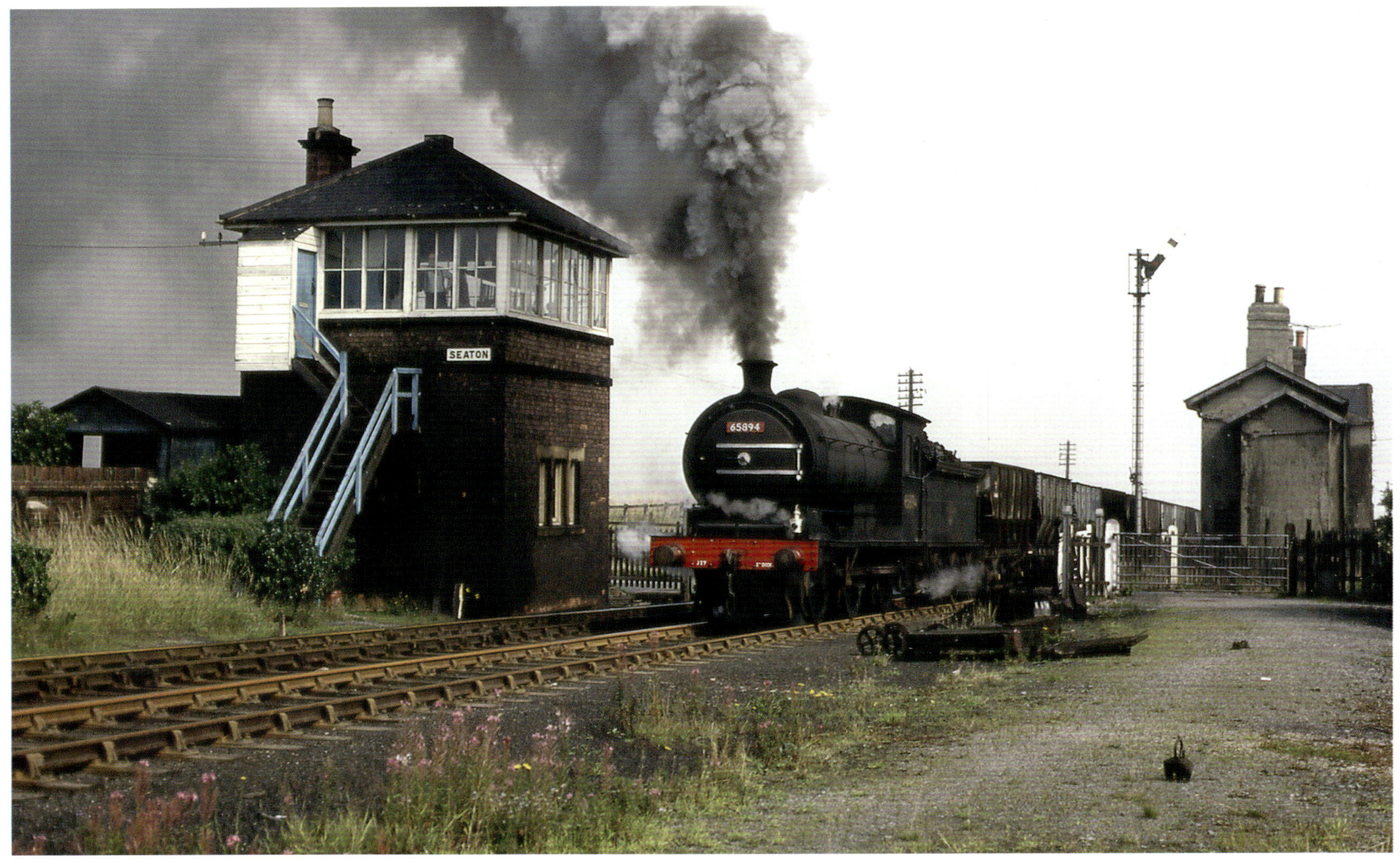

Above: Tuesday 22 August 1967. Lifting an 'up' train of coal hopper wagons past Seaton signal box, ex-NER Class P3 (LNER Class J27) 0-6-0 No. 65894 darkens the skies. Constructed at Darlington Works as the last member of the class during 1923, she would be withdrawn from service at the end of steam workings in the north-east of England in September 1967, while based at 52G Sunderland shed. Purchased by a preservation group after withdrawal, she is currently undergoing an overhaul. (Andrew G. Forsyth)

Opposite top: Thursday 28 March 1968. With Padiham Power Station in the background, ex-LMS Class 8F 2-8-0 No. 48167 is seen progressing off the line from Padiham with a short train of empty 16-ton mineral wagons, about to pass Rose Grove shed, where she was based. She had been constructed at Crewe Works during 1943 and would be withdrawn at the end of steam workings, when Rose Grove shed closed on 5 August 1968. (Andrew G. Forsyth)

Opposite bottom: Thursday 28 March 1968. In relatively clean condition, ex-LMS Class 5 4-6-0 No. 44899, fitted with a small snowplough, is seen passing Rose Grove West signal box with the shed coaling tower in the background. A product of Crewe Works during 1945, she would be withdrawn from Rose Grove shed in July 1968. (Andrew G. Forsyth)

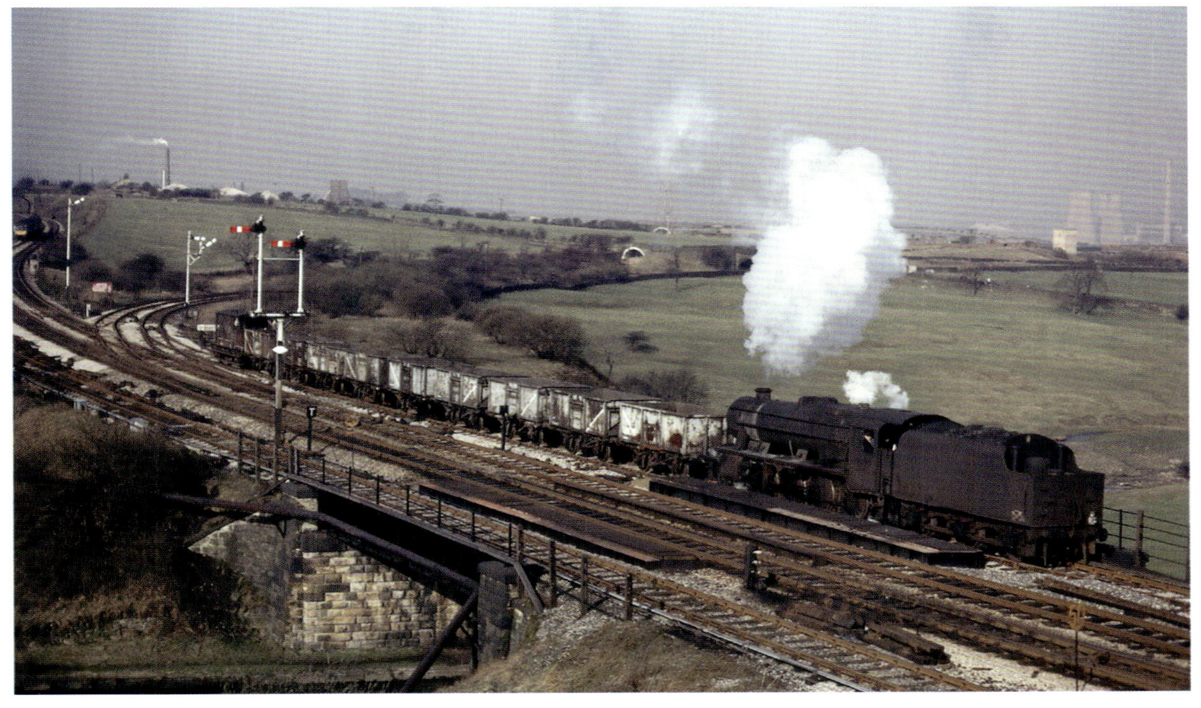

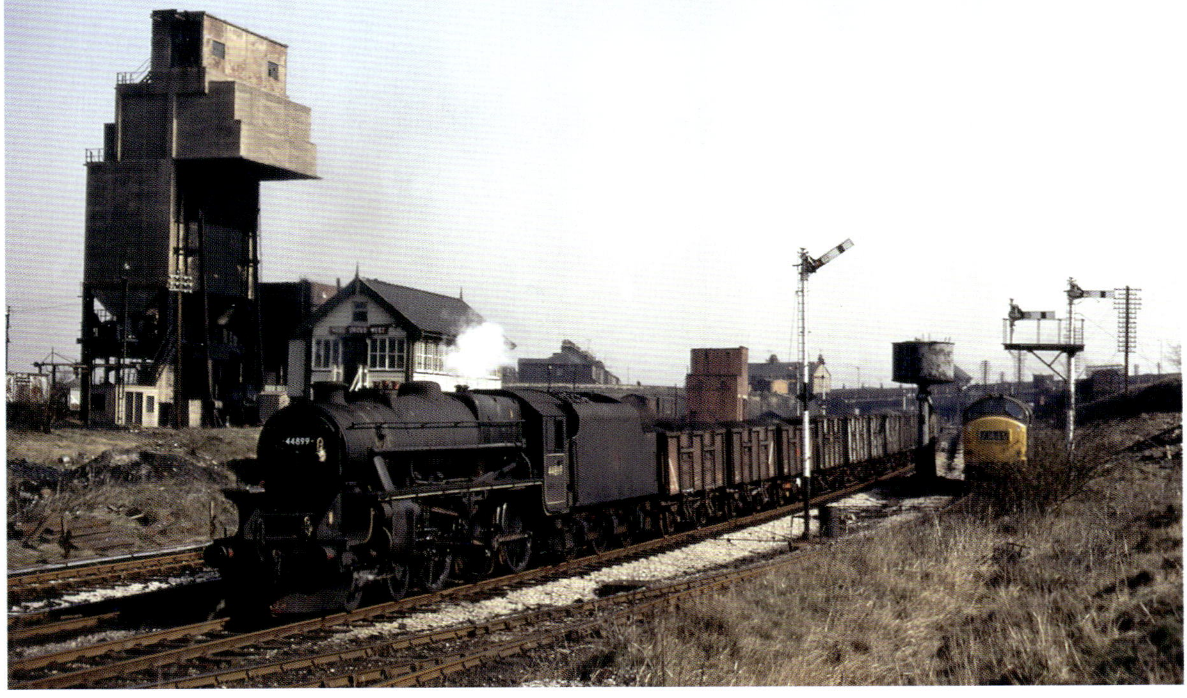

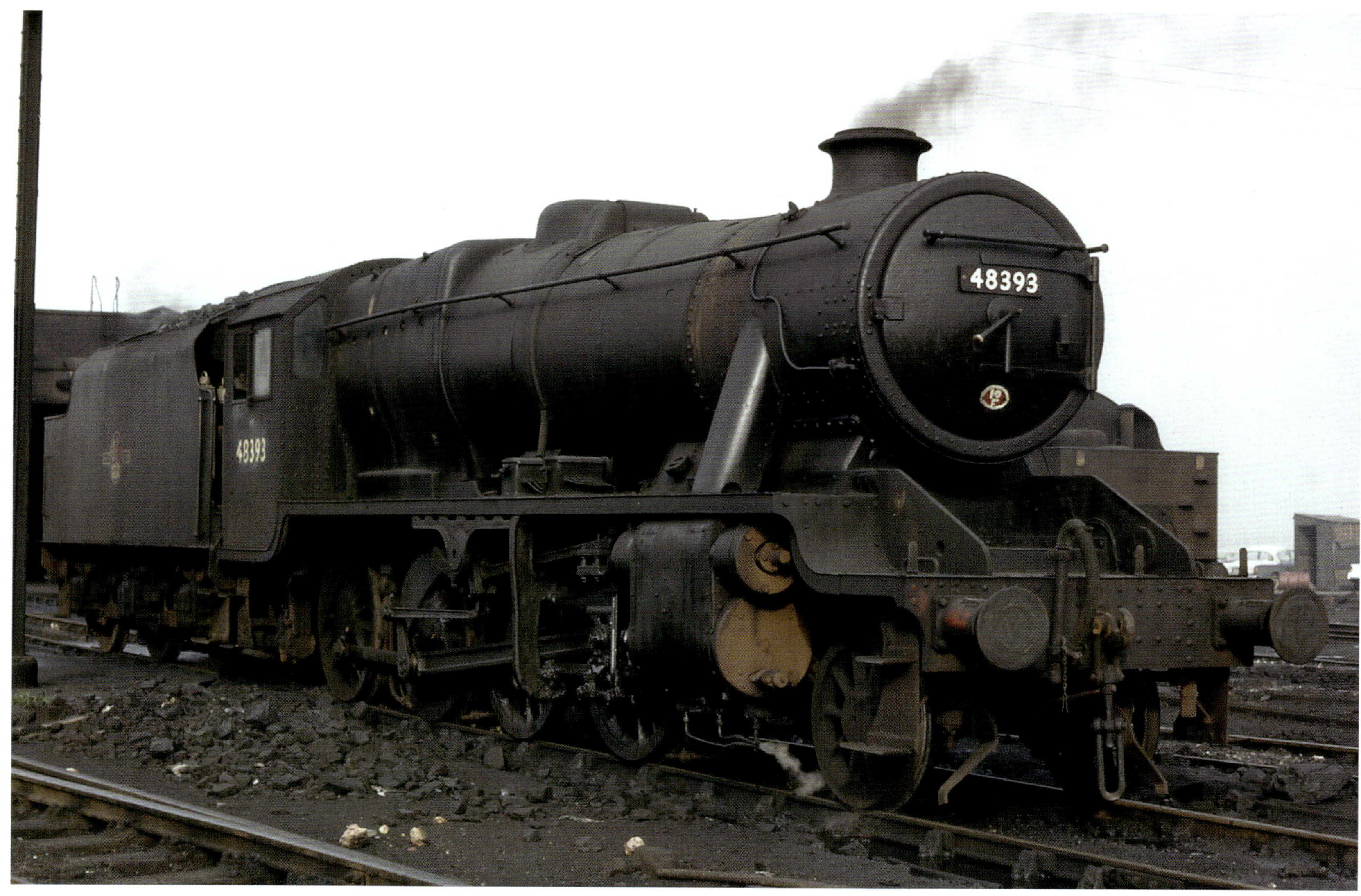

Friday 29 March 1968. This spread of photographs shows detail of ex-LMS Class 8F 2-8-0 No. 48393 while parked in the yard at Rose Grove shed. Constructed at Horwich Works in 1945 and numbered 8393 by the LMS, throughout her working life she would be allocated to a series of sheds in the Midlands, moving to Lostock Hall in 1966 and finally Rose Grove in 1967. She would be withdrawn from service when Rose Grove closed on 5 August 1968.

Above: Work stained and carrying a 10F Rose Grove shed code. (Andrew G. Forsyth)

Opposite top: The fireman's side of the locomotive in unlined black livery with her classification, 8F, on the cab side. (Andrew G. Forsyth)

Opposite bottom: The fireman's side cylinder block, slide bars, crosshead, link, combination lever and radius rod. (Andrew G. Forsyth)

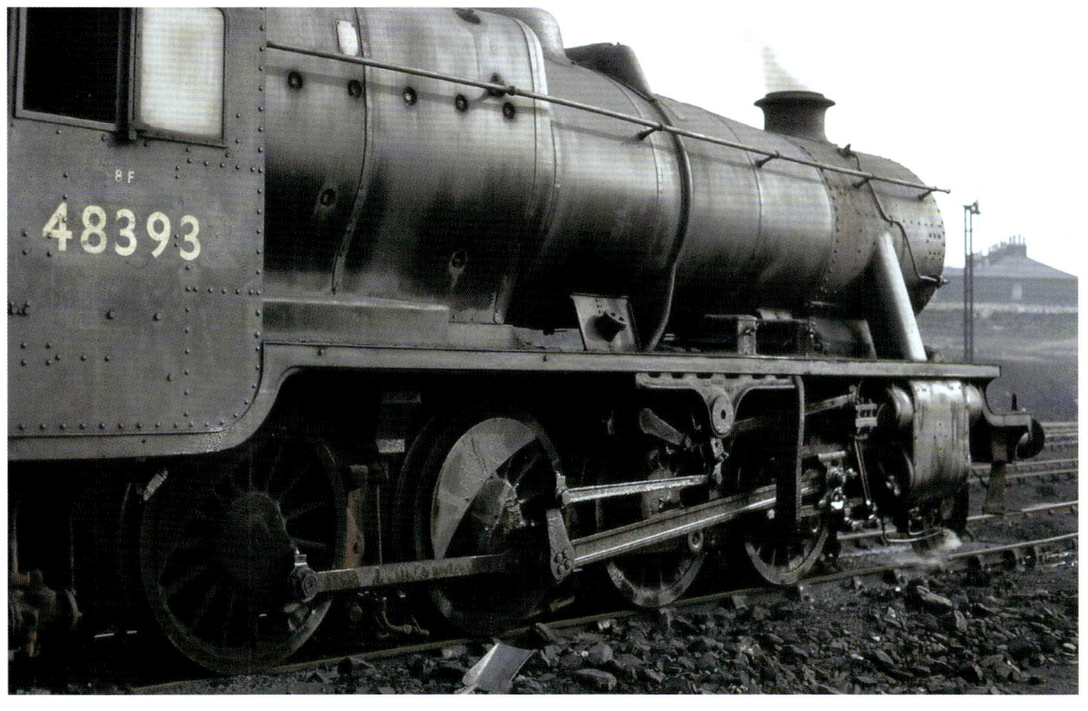

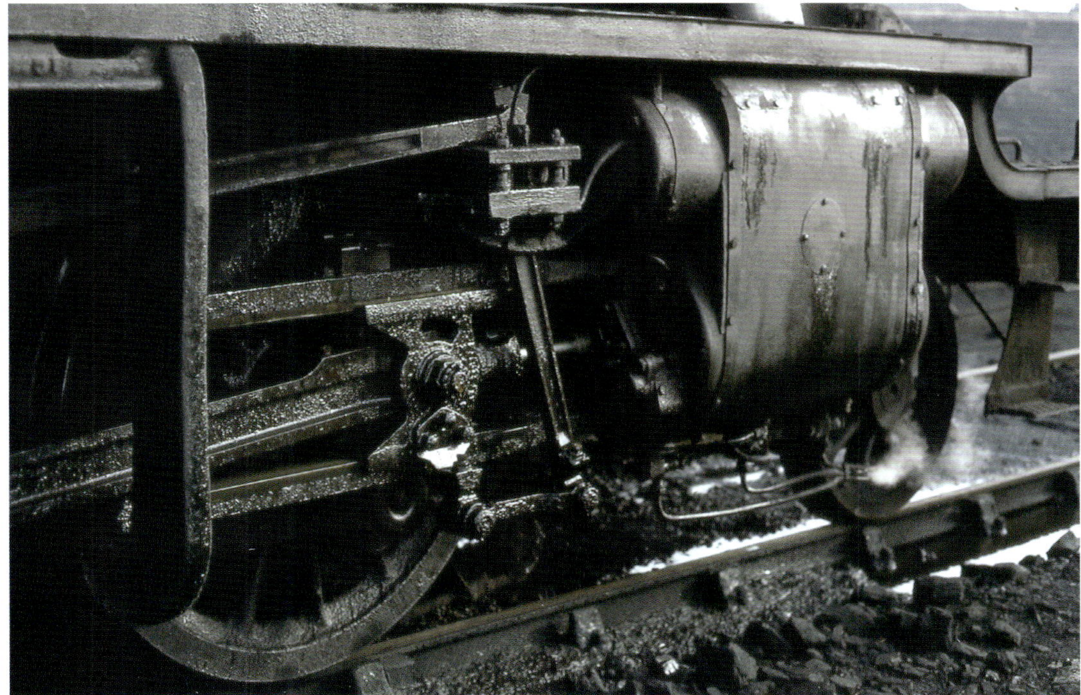

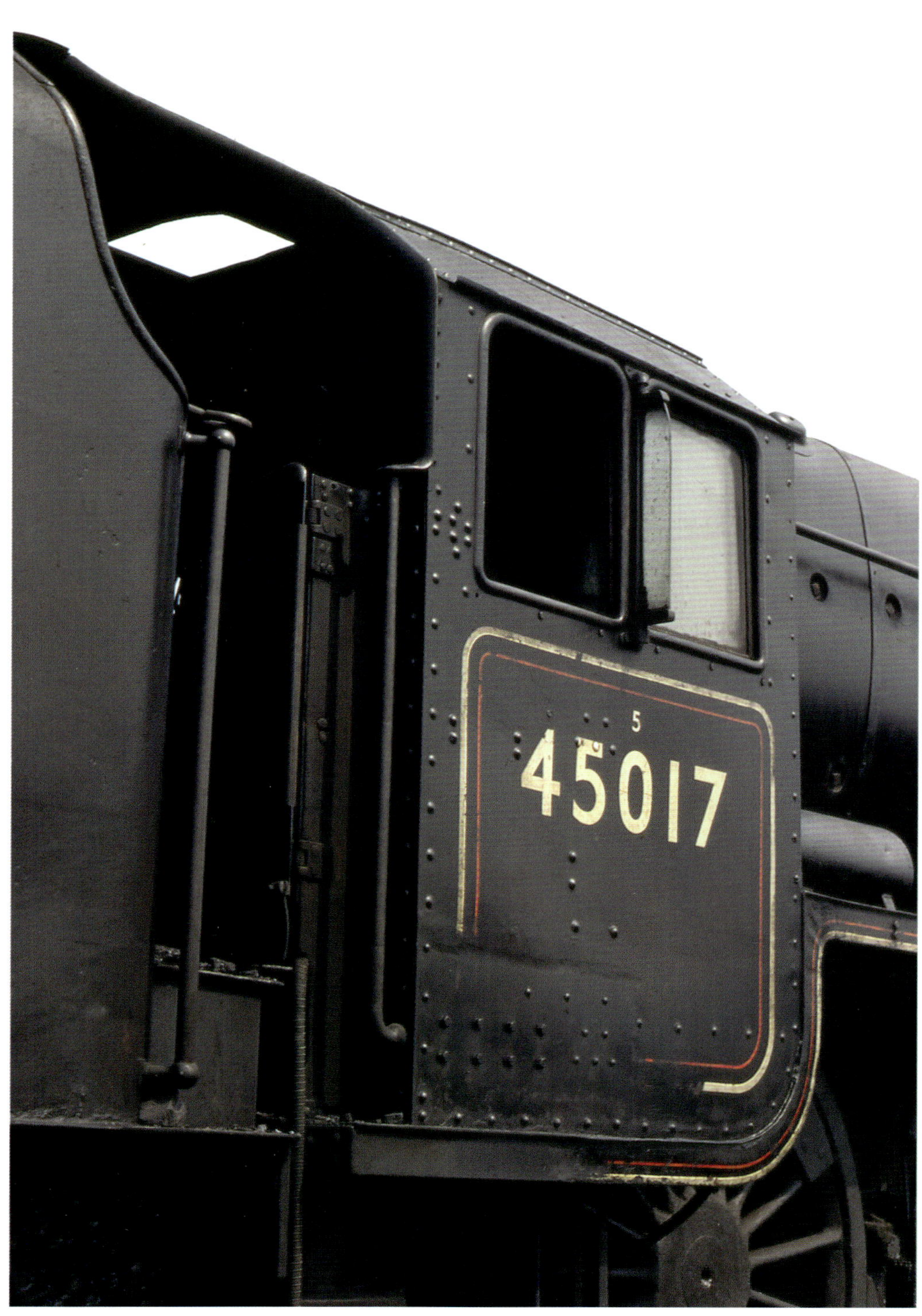

Thursday 31 July 1968. At Lostock Hall shed, ex-LMS Class 5 4-6-0 No. 45017 is looking in splendid condition in advance of her use on the SLS 'Farewell to Steam 2' tour that would take place on the following Sunday 4 August. One of the early Crewe Works-constructed examples of the class to enter service during 1935, she would end her days allocated to Carnforth shed, whose 10A code she is carrying, and be withdrawn from service while there, in August 1968.

Left: The fireman's cab side showing signs of a recent repair. (Andrew G. Forsyth)

Opposite top: The locomotive is lined out but the tender she is attached to bears no lining. (Andrew G. Forsyth)

Opposite bottom: The fireman's side cylinder block with a clean set of gear. (Andrew G. Forsyth)

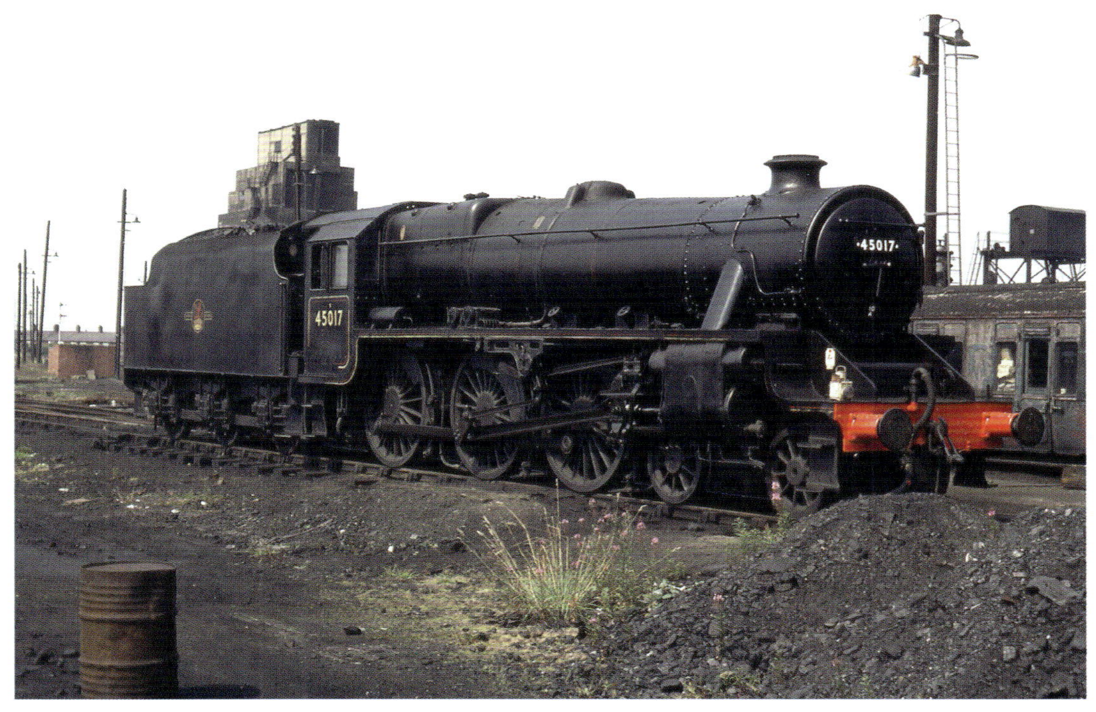

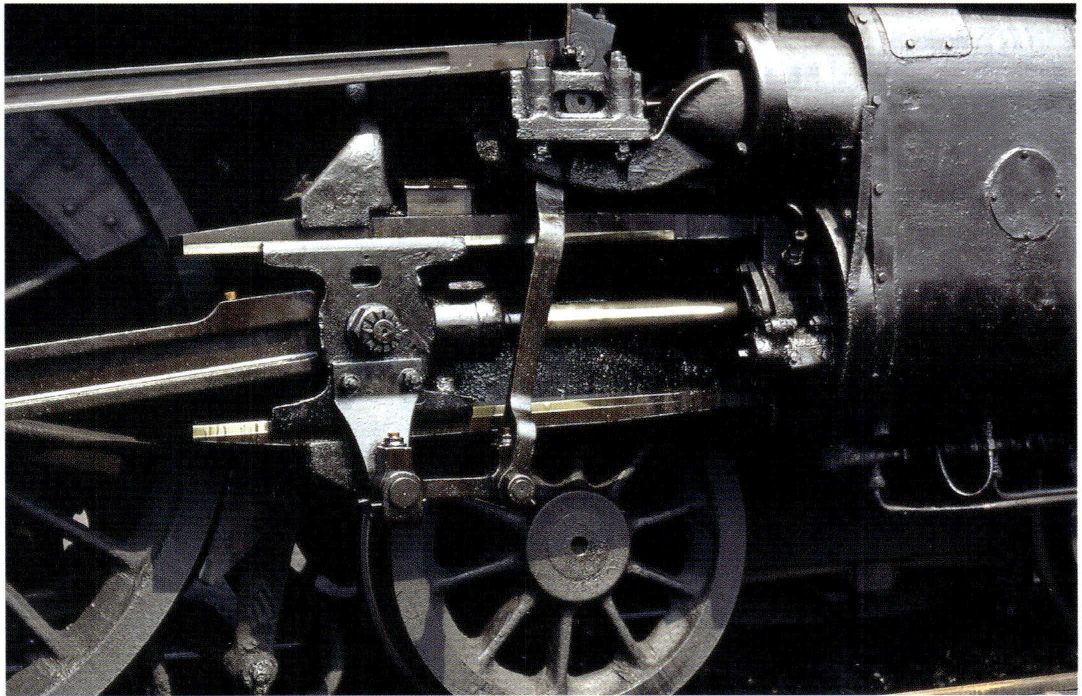

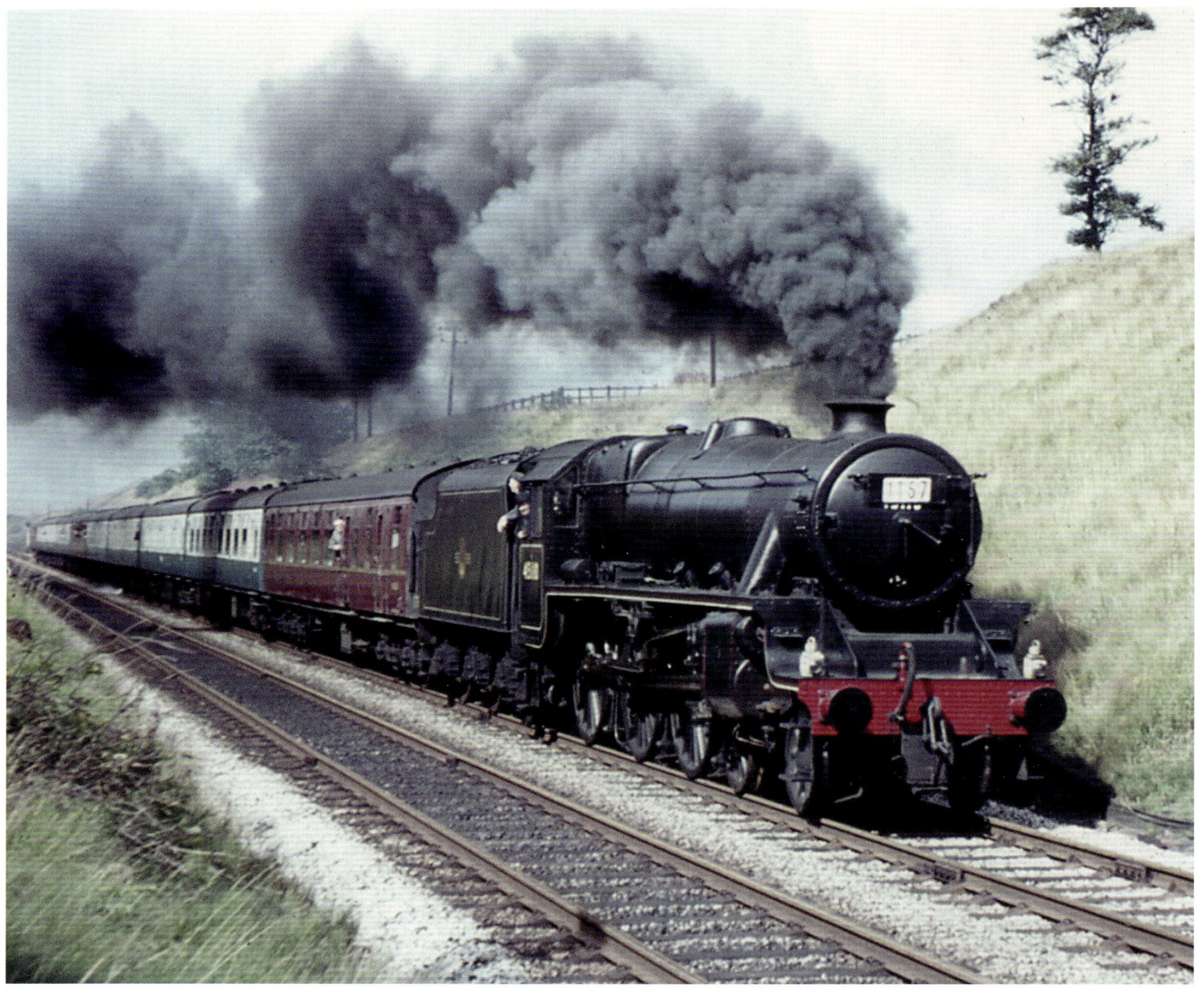

Sunday 11 August 1968. The 'Fifteen Guinea Special', organised by British Rail as the final standard-gauge passenger service hauled by steam locomotives, commenced at Liverpool Lime Street and ran to Manchester Victoria, this first section being hauled by ex-LMS Class 5 4-6-0 No. 45110. The second leg saw BR Standard Class 7P6F 4-6-2 No. 70013 *Oliver Cromwell* work the ten-coach train to Carlisle over the Settle and Carlisle route, with the return to Manchester Victoria, later that afternoon over the same route, being worked by ex-LMS Class 5 4-6-0 leading class-mate No. 44871. The final leg from Manchester to Liverpool was again in the hands of No. 45110. Seen here splendidly turned out by the Lostock Hall shed staff is No. 45110 accelerating away from Rainhill on the outbound trip. This locomotive had been the 'stand-by' substituted due to the rostered Black 5 No. 45305 being failed. Constructed by the Vulcan Foundry in 1935, No. 45110 was bought for preservation immediately after withdrawal and is currently in store at the Severn Valley Railway. (Anthony Scarsbrook)